Springer Texts in Statistics

Advisors:
George Casella   Stephen Fienberg   Ingram Olkin

Springer Texts in Statistics

Alfred: Elements of Statistics for the Life and Social Sciences
Berger: An Introduction to Probability and Stochastic Processes
Bilodeau and Brenner: Theory of Multivariate Statistics
Blom: Probability and Statistics: Theory and Applications
Brockwell and Davis: Introduction to Times Series and Forecasting, Second Edition
Chow and Teicher: Probability Theory: Independence, Interchangeability, Martingales, Third Edition
Christensen: Advanced Linear Modeling: Multivariate, Time Series, and Spatial Data; Nonparametric Regression and Response Surface Maximization, Second Edition
Christensen: Log-Linear Models and Logistic Regression, Second Edition
Christensen: Plane Answers to Complex Questions: The Theory of Linear Models, Third Edition
Creighton: A First Course in Probability Models and Statistical Inference
Davis: Statistical Methods for the Analysis of Repeated Measurements
Dean and Voss: Design and Analysis of Experiments
du Toit, Steyn, and Stumpf: Graphical Exploratory Data Analysis
Durrett: Essentials of Stochastic Processes
Edwards: Introduction to Graphical Modelling, Second Edition
Finkelstein and Levin: Statistics for Lawyers
Flury: A First Course in Multivariate Statistics
Jobson: Applied Multivariate Data Analysis, Volume I: Regression and Experimental Design
Jobson: Applied Multivariate Data Analysis, Volume II: Categorical and Multivariate Methods
Kalbfleisch: Probability and Statistical Inference, Volume I: Probability, Second Edition
Kalbfleisch: Probability and Statistical Inference, Volume II: Statistical Inference, Second Edition
Karr: Probability
Keyfitz: Applied Mathematical Demography, Second Edition
Kiefer: Introduction to Statistical Inference
Kokoska and Nevison: Statistical Tables and Formulae
Kulkarni: Modeling, Analysis, Design, and Control of Stochastic Systems
Lange: Applied Probability
Lehmann: Elements of Large-Sample Theory
Lehmann: Testing Statistical Hypotheses, Second Edition
Lehmann and Casella: Theory of Point Estimation, Second Edition
Lindman: Analysis of Variance in Experimental Design
Lindsey: Applying Generalized Linear Models

(continued after index)

Rick Durrett

Essentials of Stochastic Processes

With 15 Figures

Springer

Rick Durrett
Department of Mathematics
Cornell University
Ithaca, NY 14853-7801
USA

Editorial Board

George Casella
Biometrics Unit
Cornell University
Ithaca, NY 14853-7801
USA

Stephen Fienberg
Department of Statistics
Carnegie Mellon University
Pittsburgh, PA 15213-3890
USA

Ingram Olkin
Department of Statistics
Stanford University
Stanford, CA 94305
USA

Library of Congress Cataloging-in-Publication Data
Durrett, Richard
    Essentials of stochastic processes / Rick Durrett.
      p. cm.—(Springer texts in statistics)
    Includes bibliographical references and index.
    ISBN 0-387-98836-X (alk. paper)
    1. Stochastic processes. I. Title. II. Series.
QA274.D87   1999
519.2—dc21   99-14733

Printed on acid-free paper.

© 1999 Springer Science+Business Media, Inc.
All rights reserved. This work may not be translated or copied in whole or in part without the written permission of the publisher (Springer Science+Business Media, Inc., 233 Spring Street, New York, NY 10013, USA), except for brief excerpts in connection with reviews or scholarly analysis. Use in connection with any form of information storage and retrieval, electronic adaptation, computer software, or by similar or dissimilar methodology now known or hereafter developed is forbidden.
The use of general descriptive names, trade names, trademarks, etc., in this publication, even if the former are not especially identified, is not to be taken as a sign that such names, as understood by the Trade Marks and Merchandise Marks Act, may accordingly be used freely by anyone.

Production managed by Al an Abrams; manufacturing supervised by Thomas King.
Photocomposed pages prepared from the author's TEX files.
Printed and bound by R.R. Donnelley and Sons, Harrisonburg, VA.
Printed in the United States of America.

9 8 7 6

springer.com

Preface

How to Read This Book. The important statements and formulas are numbered sequentially in each section. Other results of more fleeting interest are denoted by $(\star)$, $(*)$, $(\oplus)$, $(\sharp)$, etc. Our most important theoretical conclusions are labeled **Theorems**. Intermediate steps toward reaching these goals are called **Lemmas**. Facts about examples are numbered but not named.

To allow readers (and instructors) to choose their own level of detail, many of the proofs begin with a nonrigorous answer to the question "**Why is this true?**" followed by a **Proof** that fills in the missing details. As it is possible to drive a car without knowing about the working of the internal combustion engine, it is also possible to apply the theory of Markov chains without knowing the details of the proofs. A little box, □, marks the ends of the proofs and of the examples. It is my personal philosophy that probability theory was developed to solve problems, so most of our effort will be spent on analyzing examples. Readers who want to master the subject will have to do more than a few of the three hundred some odd exercises.

Notes for the Instructor. Between the first undergraduate course in probability and the first graduate course that uses measure theory, there are a number of courses that teach *Stochastic Processes* to students with many different interests and with varying degrees of mathematical sophistication. For example, in the school of Operations Research and Industrial Engineering at Cornell, ORIE 361 is taken by undergraduate majors who have just completed a one-semester course in probability and statistics, and by master's students. At a higher level ORIE 650 is taken by first-year Ph.D. students who have had or are concurrently taking an advanced undergraduate class in analysis.

In order to address the needs of this wide spectrum of students, the book has been designed, like most video games, so that it can be played on various difficulty settings.

Hard. This is the easiest to describe: Do the whole book in one semester.

Normal. Omit some of the following: the proofs in Section 1.8, Section 1.3 after (3.5), the last three sections of Chapter 4 on queueing networks, spatial Poisson processes, Markov chains with infinite state space.

Easy. Drop Chapter 2 on martingales and Section 6.3 on their use to study

Brownian motion. This will make it difficult to talk about option pricing and the Black–Scholes formula but the notion of martingales could be introduced at that time.

This book began as notes I typed in the spring of 1997 as I was teaching ORIE 361 for the second time. Semyon Kruglyak read though the book when he taught that course during the summer of 1997 and made a number of valuable suggestions that were incorporated into the first official draft. That version was read by three reviewers in November of 1997. Even though they thought it was "a little bare bones," they greeted the book with enthusiasm and it was placed under contract. A second revision followed that doubled the number of pages and tripled the number of exercises. This was read by three more reviewers during May to July 1998, leading to the final version, which was prepared in December 1998 and January 1999.

Family update. It has been more than three years since my last preface. David, who was referred to then as "a.k.a. Pooh Bear" is now 12. On the threshold of adolescence, he has developed a fondness for 3D shooters like *Quake*, *Shogo Mobile Armored Division*, and *Diablo*, signing in as "Hellion" even when he plays something as innocent as the *Legend of Zelda: The Ocarina of Time*. Despite our best efforts to the contrary, Greg is turning into a math geek. His tastes run to strategy games like *Civilization II*, *Age of Empires*, and *Heroes of Might and Magic* but he also likes to show off his well developed reflexes playing *Zelda* or *Rogue's Squadron*.

Despite several threats to go back to work, Susan Myron is still a "stay at home" mother, a name that is both politically incorrect and inaccurate. We would never get by without all the energy she devotes to supervise the smaller children's homework and the largest child's chores. In between she pursues her new passions for studying the Civil War and using one-click ordering from amazon.com.

As usual this book was written in my basement office with the music cranked up loud. Hole's *Celebrity Skin* and Everclear's *So Much for the Afterglow* provided the sound track for the final version. Perhaps because of the environment in which they were created, my books seem destined to have typos. However, thanks to the efforts of Tom Liggett and his TA, Paul Jung, this second printing has 78 fewer typos than the first. Undoubtedly a few more remain. E-mail me when you find them and consult my web page for updates.

Rick Durrett

Contents

Review of Probability

1. Probabilities, Independence   1
2. Random Variables, Distributions   8
3. Expected Value, Moments   18

1 Markov Chains

1. Definitions and Examples   28
2. Multistep Transition Probabilities   34
3. Classification of States   39
4. Limit Behavior   48
5. Some Special Examples   59
6. One-Step Calculations   66
7. Infinite State Spaces   73
8. Proofs of the Convergence Theorems   81
9. Exercises   88

2 Martingales

1. Conditional Expectation   100
2. Examples of Martingales   102
3. Optional Stopping Theorem   110
4. Applications   114
5. Exercises   121

3 Poisson Processes

1. Exponential Distribution   126
2. Defining the Poisson Process   130
3. Compound Poisson Processes   137

4. Thinning and Superposition 140
  5. Conditioning 142
  6. Spatial Poisson Processes 145
  7. Exercises 152

4 Continuous-Time Markov Chains

  1. Definitions and Examples 159
  2. Computing the Transition Probability 164
  3. Limiting Behavior 169
  4. Queueing Chains 176
  5. Reversibility 181
  6. Queueing Networks 185
  7. Closed Queueing Networks 195
  8. Exercises 200

5 Renewal Theory

  1. Basic Definitions 209
  2. Laws of Large Numbers 214
  3. Applications to Queueing Theory 221
  4. Age and Residual Life 228
  5. Exercises 234

6 Brownian Motion

  1. Basic Definitions 242
  2. Markov Property, Reflection Principle 246
  3. Martingales, Hitting Times 250
  4. Option Pricing in Discrete Time 257
  5. The Black–Scholes Formula 261
  6. Exercises 265

References 271

Answers to Selected Exercises 273

Index 279

Review of Probability

Here we will review some of the basic facts usually taught in a first course in probability, concentrating on the ones that are used more than once in one of the six other chapters. This chapter may be read for review or skipped and referred to later if the need arises.

1. Probabilities, Independence

We begin with a vague but useful definition. (Here and in what follows, **boldface** indicates a word or phrase that is being defined or explained.) The term **experiment** is used to refer to any process whose outcome is not known in advance. Two simple experiments are flip a coin, and roll a die. The **sample space** associated with an experiment is the set of all possible outcomes. The sample space is usually denoted by Ω, the capital Greek letter Omega.

Example 1.1. Flip three coins. The flip of one coin has two possible outcomes, called "Heads" and "Tails," and denoted by H and T. Flipping three coins leads to $2^3 = 8$ outcomes:

$$\begin{array}{cccc} & HHT & HTT & \\ HHH & HTH & THT & TTT \\ & THH & TTH & \end{array}$$

Example 1.2. Roll two dice. The roll of one die has six possible outcomes: 1, 2, 3, 4, 5, and 6. Rolling two dice leads to $6^2 = 36$ outcomes $\{(m,n) : 1 \leq m, n \leq 6\}$.

The goal of probability theory is to compute the probability of various events of interest. Intuitively, an event is a statement about the outcome of an experiment. Formally, an **event** is a subset of the sample space. An example for flipping three coins is "two coins show Heads," or

$$A = \{HHT, HTH, THH\}$$

Review of Probability

An example for rolling two dice is "the sum is 9," or

$$B = \{(6,3), (5,4), (4,5), (3,6)\}$$

Events are just sets, so we can perform the usual operations of set theory on them. For example, if $\Omega = \{1,2,3,4,5,6\}$, $A = \{1,2,3\}$, and $B = \{2,3,4,5\}$, then the **union** $A \cup B = \{1,2,3,4,5\}$, the **intersection** $A \cap B = \{2,3\}$, and the **complement** of A, $A^c = \{4,5,6\}$. To introduce our next definition, we need one more notion: two events are **disjoint** if their intersection is the empty set, $\emptyset$. A and B are not disjoint, but if $C = \{5,6\}$, then A and C are disjoint.

A **probability** is a way of assigning numbers to events that satisfies:

(i) For any event A, $0 \leq P(A) \leq 1$.

(ii) If Ω is the sample space, then $P(\Omega) = 1$.

(iii) For a finite or infinite sequence of disjoint events $P(\cup_i A_i) = \sum_i P(A_i)$.

In words, the probability of a union of disjoint events is the sum of the probabilities of the sets. We leave the index set unspecified since it might be finite,

$$P(\cup_{i=1}^k A_i) = \sum_{i=1}^k P(A_i)$$

or it might be infinite, $P(\cup_{i=1}^\infty A_i) = \sum_{i=1}^\infty P(A_i)$.

In Examples 1.1 and 1.2, all outcomes have the same probability, so

$$P(A) = |A|/|\Omega|$$

where $|B|$ is short for the number of points in B. For a very general example, let $\Omega = \{1, 2, \ldots, n\}$; let $p_i \geq 0$ with $\sum_i p_i = 1$; and define $P(A) = \sum_{i \in A} p_i$. Two basic properties that follow immediately from the definition of a probability are

(1.1) $\qquad P(A) = 1 - P(A^c)$

(1.2) $\qquad P(B \cup C) = P(B) + P(C) - P(B \cap C)$

To illustrate their use consider the following:

Example 1.3. Roll two dice and suppose for simplicity that they are red and green. Let A = "at least one 4 appears," B = "a 4 appears on the red die," and C = "a 4 appears on the green die," so $A = B \cup C$.

Solution 1. A^c = "neither die shows a 4," which contains $5 \cdot 5 = 25$ outcomes so (1.1) implies $P(A) = 1 - 25/36 = 11/36$.

Solution 2. $P(B) = P(C) = 1/6$ while $P(B \cap C) = P(\{4, 4\}) = 1/36$, so (1.2) implies $P(A) = 1/6 + 1/6 - 1/36 = 11/36$.

Conditional probability. Suppose we are told that the event A with $P(A) > 0$ occurs. Then the sample space is reduced from Ω to A and the probability that B will occur given that A has occurred is

(1.3) $$P(B|A) = P(B \cap A)/P(A)$$

To explain this formula, note that (i) only the part of B that lies in A can possibly occur, and (ii) since the sample space is now A, we have to divide by $P(A)$ to make $P(A|A) = 1$. Multiplying on each side of (1.3) by $P(A)$ gives us the **multiplication rule**:

(1.4) $$P(A \cap B) = P(A)P(B|A)$$

Intuitively, we think of things occurring in two stages. First we see if A occurs, then we see what the probability B occurs given that A did. In many cases these two stages are visible in the problem.

Example 1.4. Suppose we draw without replacement from an urn with 6 blue balls and 4 red balls. What is the probability we will get two blue balls? Let A = blue on the first draw, and B = blue on the second draw. Clearly, $P(A) = 6/10$. After A occurs, the urn has 5 blue balls and 4 red balls, so $P(B|A) = 5/9$ and it follows from (1.4) that

$$P(A \cap B) = P(A)P(B|A) = \frac{6}{10} \cdot \frac{5}{9}$$

To see that this is the right answer notice that if we draw two balls without replacement and keep track of the order of the draws, then there are $10 \cdot 9$ outcomes, while $6 \cdot 5$ of these result in two blue balls being drawn.

EXERCISE 1.1. (a) What is the probability that two red balls will be drawn? (b) What is the probability we will draw one red and one blue (in either order)?

The multiplication rule is useful in solving a variety of problems. To illustrate its use we consider:

Example 1.5. Suppose we roll a four-sided die then flip that number of coins. What is the probability we will get exactly one Heads? Let B = we get exactly one Heads, and A_i = an i appears on the first roll. Clearly, $P(A_i) = 1/4$ for $1 \leq i \leq 4$. A little more thought gives

$$P(B|A_1) = 1/2, \quad P(B|A_2) = 2/4, \quad P(B|A_3) = 3/8, \quad P(B|A_4) = 4/16$$

Review of Probability

so breaking things down according to which A_i occurs,

$$P(B) = \sum_{i=1}^{4} P(B \cap A_i) = \sum_{i=1}^{4} P(A_i) P(B|A_i)$$
$$= \frac{1}{4}\left(\frac{1}{2} + \frac{2}{4} + \frac{3}{8} + \frac{4}{16}\right) = \frac{13}{32}$$

One can also ask the reverse question: if B occurs, what is the most likely cause? By the definition of conditional probability and the multiplication rule,

(1.5) $$P(A_i|B) = \frac{P(A_i \cap B)}{\sum_{j=1}^{4} P(A_j \cap B)} = \frac{P(A_i)P(B|A_i)}{\sum_{j=1}^{4} P(A_j)P(B|A_j)}$$

This little monster is called **Bayes' formula** but it will not see much action here.

Last but not least, two events A and B are said to be **independent** if $P(B|A) = P(B)$. In words, knowing that A occurs does not change the probability that B occurs. Using the multiplication rule this definition can be written in a more symmetric way as

(1.6) $$P(A \cap B) = P(A) \cdot P(B)$$

Example 1.6. Roll two dice and let $A =$ "the first die is 4."

Let $B_1 =$ "the second die is 2." This satisfies our intuitive notion of independence since the outcome of the first dice roll has nothing to do with that of the second. To check independence from (1.6), we note that $P(B_1) = 1/6$ while the intersection $A \cap B_1 = \{(4,2)\}$ has probability $1/36$.

$$P(A \cap B_1) = \frac{1}{36} \neq \frac{1}{6} \cdot \frac{4}{36} = P(A)P(B_1)$$

Let $B_2 =$ "the sum of the two dice is 3." The events A and B_2 are disjoint, so they cannot be independent:

$$P(A \cap B_2) = 0 < P(A)P(B_2)$$

Let $B_3 =$ "the sum of the two dice is 9." This time the occurrence of A enhances the probability of B_3, i.e., $P(B_3|A) = 1/6 > 4/36 = P(B_3)$, so the two events are not independent. To check that this claim using (1.6), we note that (1.4) implies

$$P(A \cap B_3) = P(A)P(B_3|A) > P(A)P(B_3)$$

Let $B_4 =$ "the sum of the two dice is 7." Somewhat surprisingly, A and B_4 are independent. To check this from (1.6), we note that $P(B_4) = 6/36$ and $A \cap B_4 = \{(4,3)\}$ has probability 1/36, so

$$P(A \cap B_3) = \frac{1}{36} = \frac{1}{6} \cdot \frac{6}{36} = P(A)P(B_3)$$

There are two ways of extending the definition of independence to more than two events.

$A_1, \ldots, A_n$ are said to be **pairwise independent** if for each $i \neq j$, $P(A_i \cap A_j) = P(A_i)P(A_j)$, that is, each pair is independent.

$A_1, \ldots, A_n$ are said to be **independent** if for any $1 \leq i_1 < i_2 < \ldots < i_k \leq n$ we have

$$P(A_{i_1} \cap \ldots \cap A_{i_k}) = P(A_{i_1}) \cdots P(A_{i_k})$$

If we flip n coins and let $A_i =$ "the ith coin shows Heads," then the A_i are independent since $P(A_i) = 1/2$ and for any choice of indices $1 \leq i_1 < i_2 < \ldots < i_k \leq n$ we have $P(A_{i_1} \cap \ldots \cap A_{i_k}) = 1/2^k$. Our next example shows that events can be pairwise independent but not independent.

Example 1.7. Flip three coins. Let $A =$ "the first and second coins are the same," $B =$ "the second and third coins are the same," and $C =$ "the third and first coins are the same." Clearly $P(A) = P(B) = P(C) = 1/2$. The intersection of any two of these events is

$$A \cap B = B \cap C = C \cap A = \{HHH, TTT\}$$

an event of probability 1/4. From this it follows that

$$P(A \cap B) = \frac{1}{4} = \frac{1}{2} \cdot \frac{1}{2} = P(A)P(B)$$

i.e., A and B are independent. Similarly, B and C are independent and C and A are independent; so A, B, and C are pairwise independent. The three events A, B, and C are not independent, however, since $A \cap B \cap C = \{HHH, TTT\}$ and hence

$$P(A \cap B \cap C) = \frac{1}{4} \neq \left(\frac{1}{2}\right)^3 = P(A)P(B)P(C)$$

The last example is somewhat unusual. However, the moral of the story is that to show several events are independent, you have to check more than just that each pair is independent.

Review of Probability

EXERCISES

1.2. Suppose we pick a letter at random from the word TENNESSEE. What is the sample space Ω and what probabilities should be assigned to the outcomes?

1.3. Suppose we roll a red die and a green die. What is the probability the number on the red die is larger ($>$) than the number on the green die?

1.4. Two dice are rolled. What is the probability that (a) the two numbers will differ by 1 or less, (b) the maximum of the two numbers will be 5 or larger, (c) the maximum of the two numbers is $= m$?

1.5. In the World Series, two teams play until one team has won four games. Suppose that the outcome of each game is determined by flipping a coin. What is the probability that the World Series will last (a) four games, (b) five games, (c) six games, (d) seven games?

1.6. In a group of students, 25% smoke cigarettes (event C), 60% drink alcohol (event A), and 15% do both. (a) What fraction of students have at least one of these bad habits? (b) Are A and C independent?

1.7. In a group of 320 high school graduates, only 160 went to college, but 100 of the 170 men did. (a) How many women did not go to college? (b) Let C stand for college and M stand for men. Are the events C and M independent? (c) Compute $P(C|M)$. (d) Compute $P(M|C)$.

1.8. Suppose $\Omega = \{a, b, c\}$, $P(\{a,b\}) = 0.7$, and $P(\{b,c\}) = 0.6$. Compute the probabilities of $\{a\}$, $\{b\}$, and $\{c\}$.

1.9. Suppose we draw two cards out of a deck of 52. Let $A = $ " The first card is an Ace," $B = $ "The second card is a spade." Are A and B independent?

1.10. A family has several children, each of whom is independently a boy or a girl with probability 1/2. Let $A = $ "There is at most one girl," $B = $ "The family has children of both sexes." (a) Are A and B independent if the family has three children? (b) four children? (c) $m \geq 5$ children?

1.11. Roll two dice. Let $A = $ "The first die is odd," $B = $ "The second die is odd," and $C = $ "The sum is odd." Show that these events are pairwise independent but not independent.

1.12. Show that A and B^c are independent if A and B are.

1.13. Give an example to show that we may have (i) A and B are independent, (ii) A and C are independent, but (iii) A and $B \cup C$ are not independent. Can this happen if B and C are disjoint?

1.14. Two students, Alice and Betty, are registered for a statistics class. Alice attends 80% of the time, Betty 60% of the time, and their absences are independent. On a given day, what is the probability (a) at least one of these students is in class, (b) exactly one of them is there, (c) Alice is there given that only student is there?

1.15. Let A and B be two independent events with $P(A) = 0.4$ and $P(A \cup B) = 0.7$. What is $P(B)$?

1.16. A friend flips two coins and tells you that at least one is Heads. Given this information, what is the probability that the first coin is Heads?

1.17. A friend rolls two dice and tells you that there is at least one 6. What is the probability the sum is at least 9?

1.18. Suppose 60% of the people subscribe to newspaper A, 40% to newspaper B, and 30% to both. If we pick a person at random who subscribes to at least one newspaper, what is the probability she subscribes to newspaper A?

1.19. Suppose that the probability a married man votes is 0.45, the probability a married woman votes is 0.4, and the probability a woman votes given that her husband does is 0.6. What is the probability (a) both vote, (b) a man votes given that his wife does?

1.20. Two events have $P(A) = 1/4$, $P(B|A) = 1/2$, and $P(A|B) = 1/3$. Compute (a) $P(A \cap B)$, (b) $P(B)$, (c) $P(A \cup B)$.

1.21. The population of Cyprus is 70% Greek and 30% Turkish. Twenty % of the Greeks and 10% of the Turks speak English. What fraction of the people of Cyprus speak English?

1.22. You are going to meet a friend at the airport. Your experience tells you that the plane is late 70% of the time when it rains, but is late only 20% of the time when it does not rain. The weather forecast that morning calls for a 40% chance of rain. What is the probability the plane will be late?

1.23. A student is taking a multiple-choice test in which each question has four possible answers. She knows the answers to 50% of the questions, can narrow the choices down to two 30% of the time, and does not know anything about 20% of the questions. What is the probability she will correctly answer a question chosen at random from the test?

1.24. A student is taking a multiple-choice test in which each question has four possible answers. She knows the answers to 5 of the questions, can narrow the choices down to two in 3 cases, and does not know anything about 2 of the questions. What is the probability she will correctly answer (a) 10, (b) 9, (c) 8, (d) 7, (e) 6, (f) 5 questions?

1.25. A group of n people is going to sit in a row of n chairs. By considering the chair Mr. Jones sits in, compute the probability Mr. Jones will sit next to Miss Smith.

1.26. Suppose we roll two dice twice. What is the probability we get the same sum on both throws?

1.27. In the game of craps when "the shooters point is 5," his objective is to throw a sum of 5 with two dice before a sum of 7 appears. There are two ways to compute the probability p he will succeed. (a) Consider the outcome of the first roll to conclude that $p = 4/36 + (26/36)p$ and then solve to find p. (b) Let S_k be the event that the sum is k and argue that $p = P(S_5)/P(S_5 \cap S_7)$.

1.28. Two boys take turns throwing darts at a target. Al throws first and hits with probability $1/5$. Bob throws second and hits with probability b. (a) Suppose $b = 1/3$. What is the probability Al will hit the target before Bob does? (b) What value of b makes each boy have a probability $1/2$ of winning?

2. Random Variables, Distributions

Formally, a **random variable** is a real-valued function defined on the sample space. However, in most cases the sample space is usually not visible, so we typically describe the random variables by giving their distributions. In the **discrete case** where the random variable can take on a finite or countably infinite set of values this is usually done using the **probability function**. That is, we give $P(X = x)$ for each value of x for which $P(X = x) > 0$.

Example 2.1. Binomial distribution. If we perform an experiment n times and on each trial there is a probability p of success, then the number of successes S_n has

$$P(S_n = k) = \binom{n}{k} p^k (1-p)^{n-k} \quad \text{for } k = 0, \ldots, n$$

In words, S_n has a binomial distribution with parameters n and p, a phrase we will abbreviate as $S_n = \text{binomial}(n, p)$.

Example 2.2. Geometric distribution. If we repeat an experiment with probability p of success until a success occurs, then the number of trials required, N, has

$$P(N = n) = (1-p)^{n-1} p \quad \text{for } n = 1, 2, \ldots$$

In words, N has a geometric distribution with parameter p, a phrase we will abbreviate as $N = \text{geometric}(p)$.

Example 2.3. Poisson distribution. X is said to have a Poisson distribution with parameter $\lambda > 0$, or $X = \text{Poisson}(\lambda)$ if

$$P(X = k) = e^{-\lambda} \frac{\lambda^k}{k!} \quad \text{for } k = 0, 1, 2, \ldots$$

To see that this is a probability function we recall

(2.1) $$e^x = \sum_{k=0}^{\infty} \frac{x^k}{k!}$$

so the proposed probabilities are nonnegative and sum to 1. Our next result will explain why the Poisson distribution arises in a number of situations.

(2.2) **Poisson approximation to the binomial.** *Suppose S_n has a binomial distribution with parameters n and p_n. If $p_n \to 0$ and $np_n \to \lambda$ as $n \to \infty$, then*

$$P(S_n = k) \to e^{-\lambda} \frac{\lambda^k}{k!}$$

In words, if we have a large number of independent events with a small probability, then the number that occur has approximately a Poisson distribution. We will prove this result in Section 3.2.

In many situations random variables can take any value on the real line or in a certain subset of the real line. For concrete examples, consider the height or weight of a person chosen at random or the time it takes a person to drive from Los Angeles to San Francisco. A random variable X is said to have a **continuous distribution** with **density function** f if for all $a \leq b$ we have

(2.3) $$P(a \leq X \leq b) = \int_a^b f(x)\, dx$$

Geometrically, $P(a \leq X \leq b)$ is the area under the curve f between a and b.

In order for $P(a \leq X \leq b)$ to be nonnegative for all a and b and for $P(-\infty < X < \infty) = 1$ we must have

(2.4) $$f(x) \geq 0 \quad \text{and} \quad \int_{-\infty}^{\infty} f(x)\, dx = 1$$

Any function f that satisfies (2.4) is said to be a **density function**. We will now define three of the most important density functions.

Review of Probability

Example 2.4. The uniform distribution on (a,b).

$$f(x) = \begin{cases} 1/(b-a) & a < x < b \\ 0 & \text{otherwise} \end{cases}$$

The idea here is that we are picking a value "at random" from (a,b). That is, values outside the interval are impossible, and all those inside have the same probability density. Note that the last property implies $f(x) = c$ for $a < x < b$. In this case the integral is $c(b-a)$, so we must pick $c = 1/(b-a)$.

Example 2.5. The exponential distribution.

$$f(x) = \begin{cases} \lambda e^{-\lambda x} & x \geq 0 \\ 0 & \text{otherwise} \end{cases}$$

Here $\lambda > 0$ is a parameter. To check that this is a density function, we note that

$$\int_0^\infty \lambda e^{-\lambda x}\, dx = -e^{-\lambda x}\Big|_0^\infty = 0 - (-1) = 1$$

Exponentially distributed random variables often come up as waiting times between events, for example, the arrival times of customers at a bank or ice cream shop. This will be discussed in great detail in Sections 3.1 and 3.2. Following the trend in Examples 2.1–2.3, we will sometimes indicate that X has an exponential distribution with parameter λ by writing $X = \text{exponential}(\lambda)$.

In a first course in probability, the next example is the star of the show. However, it will have only a minor role here.

Example 2.6. The standard normal distribution.

$$f(x) = (2\pi)^{-1/2} e^{-x^2/2}$$

Since there is no closed form expression for the antiderivative of f, it takes some ingenuity to check that this is a probability density. Those details are not important here so we will ignore them.

Any random variable (discrete, continuous, or in between) has a **distribution function** defined by $F(x) = P(X \leq x)$. If X has a density function $f(x)$ then

$$F(x) = P(-\infty < X \leq x) = \int_{-\infty}^{x} f(y)\, dy$$

That is, F is an antiderivative of f.

One of the reasons for computing the distribution function is explained by the next formula. If $a < b$, then $\{X \le b\} = \{X \le a\} \cup \{a < X \le b\}$ with the two sets on the right-hand side disjoint so

$$P(X \le b) = P(X \le a) + P(a < X \le b)$$

or, rearranging,

(2.5) $\qquad P(a < X \le b) = P(X \le b) - P(X \le a) = F(b) - F(a)$

The last formula is valid for any random variable. When X has density function f, it says that

$$\int_a^b f(x)\, dx = F(b) - F(a)$$

i.e., the integral can be evaluated by taking the difference of the antiderivative at the two endpoints.

To see what distribution functions look like, and to explain the use of (2.5), we return to our examples.

Example 2.7. The uniform distribution. $f(x) = 1/(b-a)$ for $a < x < b$.

$$F(x) = \begin{cases} 0 & x \le a \\ (x-a)/(b-a) & a \le x \le b \\ 1 & x \ge b \end{cases}$$

To check this, note that $P(a < X < b) = 1$ so $P(X \le x) = 1$ when $x \ge b$ and $P(X \le x) = 0$ when $x \le a$. For $a \le x \le b$ we compute

$$P(X \le x) = \int_{-\infty}^x f(y)\, dy = \int_a^x \frac{1}{b-a}\, dy = \frac{x-a}{b-a}$$

In the most important special case $a = 0$, $b = 1$ we have $F(x) = x$ for $0 \le x \le 1$.

Example 2.8. The exponential distribution. $f(x) = \lambda e^{-\lambda x}$ for $x \ge 0$.

$$F(x) = \begin{cases} 0 & x \le 0 \\ 1 - e^{-\lambda x} & x \ge 0 \end{cases}$$

The first line of the answer is easy to see. Since $P(X > 0) = 1$, we have $P(X \le x) = 0$ for $x \le 0$. For $x \ge 0$ we compute

$$P(X \le x) = \int_0^x \lambda e^{-\lambda y}\, dy = -e^{-\lambda y}\Big|_0^x = 1 - e^{-\lambda x}$$

Suppose X has an exponential distribution with parameter λ. If $t \geq 0$, then $P(X > t) = 1 - P(X \leq t) = 1 - F(t) = e^{-\lambda t}$, so if $s \geq 0$, then

$$P(T > t + s | T > t) = \frac{P(T > t + s)}{P(T > t)} = \frac{e^{-\lambda(t+s)}}{e^{-\lambda t}} = e^{-\lambda s} = P(T > s)$$

This is the **lack of memory property** of the exponential distribution. Given that you have been waiting t units of time, the probability you must wait an additional s units of time is the same as if you had not been waiting at all. We will explore this property in depth in Section 3.1.

In many situations we need to know the relationship between several random variables $X_1, \ldots, X_n$. If the X_i are discrete random variables then this is easy, we simply give the probability function that specifies the value of

$$P(X_1 = x_1, \ldots, X_n = x_n)$$

whenever this is positive. When the individual random variables have continuous distributions this is described by giving the **joint density function** which has the interpretation that

$$P((X_1, \ldots, X_n) \in A) = \int \cdots \int_A f(x_1, \ldots, x_n) \, dx_1 \ldots dx_n$$

By analogy with (2.4) we must require that $f(x_1, \ldots, x_n) \geq 0$ and

$$\int \cdots \int f(x_1, \ldots, x_n) \, dx_1 \ldots dx_n = 1$$

Having introduced the joint distribution of n random variables, we will for simplicity restrict our attention for the rest of the section to $n = 2$, where will typically write $X_1 = X$ and $X_2 = Y$. The first question we will confront is: "Given the joint distribution of (X, Y), how do we recover the distributions of X and Y?" In the discrete case this is easy. The **marginal distributions** of X and Y are given by

(2.6)
$$P(X = x) = \sum_y P(X = x, Y = y)$$
$$P(Y = y) = \sum_x P(X = x, Y = y)$$

To explain the first formula in words, if $X = x$, then Y will take on some value y, so to find $P(X = x)$ we sum the probabilities of the disjoint events $\{X = x, Y = y\}$ over all the values of y.

Formula (2.6) generalizes in a straightforward way to continuous distributions: we replace the sum by an integral and the probability functions by density functions. To make the analogy more apparent we will introduce the following:

Notational convention. If X has a continuous distribution we will write $P(X = x)$ for its density function, keeping in mind that in the continuous case the probability that X is exactly equal to x, $\{X = x\}$ is equal to 0. Similarly, we will write $P(X = x, Y = y)$ for the joint density of X and Y. The advantages (and perils) of this compared to the traditional approach of writing $f_X(x)$ and $f_{XY}(x, y)$ should already be clear in the next formula.

In the continuous case the **marginal densities** of X and Y are given by

(2.7)
$$P(X = x) = \int P(X = x, Y = y)\, dy$$
$$P(Y = y) = \int P(X = x, Y = y)\, dx$$

The verbal explanation of the first formula is similar to that of the discrete case: if $X = x$, then Y will take on some value y, so to find $P(X = x)$ we integrate the joint density $P(X = x, Y = y)$ over all possible values of y. The point of our notation is to bring out the analogy with the discrete case. People who are disturbed by it can simply rewrite the equations in the less intuitive form

(2.7')
$$f_X(x) = \int f_{XY}(x, y)\, dy$$
$$f_Y(y) = \int f_{XY}(x, y)\, dx$$

Two random variables are said to be **independent** if for any two sets A and B we have

(2.8) $$P(X \in A, Y \in B) = P(X \in A)P(Y \in B)$$

In the discrete case, (2.8) is equivalent to

(2.9) $$P(X = x, Y = y) = P(X = x)P(Y = y)$$

for all x and y. With our notation the condition for independence is exactly the same in the continuous case, though in that situation we must remember that the formula says that the joint distribution is the product of the marginal densities. In the traditional notation, the condition for continuous random variables is

(2.10) $$f_{XY}(x,y) = f_X(x)f_Y(y).$$

The notions of independence extend in a straightforward way to n random variables. Using our notation that combines the discrete and the continuous case $X_1, \ldots X_n$ are independent if

(2.11) $$P(X_1 = x_1, \ldots, X_n = x_n) = P(X_1 = x_1) \cdots P(X_n = x_n)$$

That is, if the joint probability or probability density is the product of the marginals. Two important consequences of independence are

(2.12) **Theorem.** If $X_1, \ldots X_n$ are independent, then
$$E(X_1 \cdots X_n) = EX_1 \cdots EX_n$$

(2.13) **Theorem.** If $X_1, \ldots X_n$ are independent and $n_1 < \ldots < n_k \leq n$, then
$$h_1(X_1, \ldots X_{n_1}), h_2(X_{n_1+1}, \ldots X_{n_2}), \ldots h_k(X_{n_{k-1}+1}, \ldots X_{n_k})$$
are independent.

In words, the second result says that functions of disjoint sets of independent random variables are independent. Of course, (2.13) can be combined with (2.12) to conclude that expectations of such products are products of the individual expected values.

Our last topic in this section is the distribution of $X + Y$ when X and Y are independent. In the discrete case this is easy:

(2.14)
$$P(X+Y = z) = \sum_x P(X = x, Y = z - x)$$
$$= \sum_x P(X = x)P(Y = z - x)$$

To see the first equality, note that if the sum is z then X must take on some value x and Y must be $z-x$. The first equality is valid for any random variables. The second holds since we have supposed X and Y are independent.

Example 2.9. If $X = \text{Poisson}(\lambda)$ and $Y = \text{Poisson}(\mu)$ are independent then $X + Y = \text{Poisson}(\lambda + \mu)$.

Proof. A computationally simpler proof will be given in Example 3.8. However, for the moment we will just slug it out with the definition

$$P(X+Y=n) = \sum_{m=0}^{n} P(X=m)P(Y=n-m)$$

$$= \sum_{m=0}^{n} e^{-\lambda}\frac{\lambda^m}{m!} \cdot e^{-\mu}\frac{\mu^{n-m}}{(n-m)!}$$

$$= e^{-(\lambda+\mu)}\frac{1}{n!} \sum_{m=0}^{n} \binom{n}{m}\lambda^m \mu^{n-m}$$

Recalling that the **Binomial theorem** says

(2.15) $$\sum_{m=0}^{n} \binom{n}{m}\lambda^m \mu^{n-m} = (\lambda+\mu)^n$$

the desired result follows. □

Example 2.10. If $X = $ binomial(n,p) and $Y = $ binomial(m,p) are independent, then $X+Y = $ binomial$(n+m,p)$.

Proof by direct computation.

$$P(X+Y=i) = \sum_{j=0}^{i} \binom{n}{j}p^j(1-p)^{n-j} \cdot \binom{m}{i-j}p^{i-j}(1-p)^{m-i+j}$$

$$= p^i(1-p)^{n+m-i} \sum_{j=0}^{i} \binom{n}{j} \cdot \binom{m}{i-j}$$

$$= \binom{n+m}{i}p^i(1-p)^{n+m-i}$$

The last equality follows from the fact that if we pick i individuals from a group of n boys and m girls, which can be done in $\binom{n+m}{i}$ ways, then we must have j boys and $i-j$ girls for some j with $0 \le j \le i$. □

Much easier proof. Consider a sequence of $n+m$ independent trials. Let X be the number of successes in the first n trials and Y be the number of successes in the last m. By (2.13), X and Y independent. Clearly their sum is binomial(n,p). □

Formula (2.14) generalizes in the usual way to continuous distributions: regard the probabilities as density functions and replace the sum by an integral.

(2.16) $$P(X+Y=z) = \int P(X=x)P(Y=z-x)\,dx$$

Example 2.11. Let U and V be independent and uniform on $(0,1)$. Compute the density function for $U+V$.

Solution. If $U+V = x$ with $0 \le x \le 1$, then we must have $U \le x$ so that $V \ge 0$. Recalling that we must also have $U \ge 0$

$$P(U+V=x) = \int_0^x 1 \cdot 1\,du = x \quad \text{when } 0 \le x \le 1$$

If $U+V = x$ with $1 \le x \le 2$, then we must have $U \ge x-1$ so that $V \le 1$. Recalling that we must also have $U \le 1$,

$$P(U+V=x) = \int_{x-1}^1 1 \cdot 1\,du = 2-x \quad \text{when } 1 \le x \le 2$$

Combining the two formulas we see that the density function for the sum is triangular. It starts at 0 at 0, increases linearly with rate 1 until it reaches the value of 1 at $x = 1$, then it decreases linearly back to 0 at $x = 2$. □

The result for sums of our other favorite continuous distribution, the exponential, is more complicated.

Example 2.12. Let $t_1, \ldots, t_n$ be random variables that are independent and have an exponential distribution with parameter λ. Then $T_n = t_1 + \cdots + t_n$ has the **gamma(n, λ) density function**

(2.17) $$f(x) = \frac{\lambda^n x^{n-1}}{(n-1)!} e^{-\lambda x}$$

This will be proved as (1.11) in Chapter 3, so we will not give the details here.

EXERCISES

2.1. Suppose we roll two dice and let X and Y be the two numbers that appear. Find the probability function for (a) $\max\{X,Y\}$, (b) $|X-Y|$.

2.2. Suppose we roll three tetrahedral dice that have 1, 2, 3, and 4 on their four sides. Find the probability function for the sum of (a) two dice, (b) three dice.

2.3. Suppose we draw three balls out of an urn that contains fifteen balls numbered from 1 to 15. Let X be the largest number drawn. Find the probability function for X.

2.4. Suppose X has a Poisson distribution with parameter λ. Compute the ratio $P(X = i)/P(X = (i-1))$ for $i \geq 1$. This recursion is useful for computing Poisson probabilities on a computer and it also shows us that $P(X = i) > P(X = i-1)$ if and only if $\lambda > i$.

2.5. Suppose $X = \text{binomial}(n, p)$. Compute $P(X = i)/P(X = (i-1))$ for $1 \leq i \leq n$. What value of k maximizes $P(X = k)$ when X is binomial with parameters n and p?

2.6. Compare the Poisson approximation with the exact binomial probabilities of 1 success when $n = 20$, $p = 0.1$.

2.7. Compare the Poisson approximation with the exact binomial probabilities of no success when (a) $n = 10$, $p = 0.1$, (b) $n = 50$, $p = 0.02$.

2.8. The probability of a three of a kind in poker is approximately 1/50. Use the Poisson approximation to estimate the probability you will get at least one three of a kind if you play 20 hands of poker.

2.9. Suppose 1% of a certain brand of Christmas lights is defective. Use the Poisson approximation to compute the probability that in a box of 25 there will be at most one defective bulb.

2.10. $F(x) = 3x^2 - 2x^3$ for $0 < x < 1$ (with $F(x) = 0$ if $x \leq 0$ and $F(x) = 1$ if $x \geq 1$) defines a distribution function. Find the corresponding density function.

2.11. Let $F(x) = e^{-1/x}$ for $x \geq 0$, $F(x) = 0$ for $x \leq 0$. Is F a distribution function? If so, find its density function.

2.12. Consider $f(x) = c(1 - x^2)$ for $-1 < x < 1$, 0 otherwise. What value of c should we take to make f a density function?

Factoid. If m is a number with $P(X \leq m) = 1/2$, then m is said to be a **median** of the distribution.

2.13. Suppose X has density function $f(x) = x/2$ for $0 < x < 2$, 0 otherwise. Find (a) the distribution function, (b) $P(X < 1)$, (c) $P(X > 3/2)$, (d) the value of m so that $P(X \leq m) = 1/2$.

2.14. Suppose X has density function $f(x) = 4x^3$ for $0 < x < 1$, 0 otherwise. Find (a) the distribution function, (b) $P(X < 1/2)$, (c) $P(X > 2/3)$, (d) the value of m so that $P(X \leq m) = 1/2$.

2.15. Suppose X has an exponential distribution with parameter λ. (a) Find $P(X > 2/\lambda)$ and (b) the value of m so that $P(X \leq m) = 1/2$.

2.16. Suppose X and Y are independent and take the values 1, 2, 3, and 4 with probabilities 0.1, 0.2, 0.3, 0.4. Find the probability function for the sum $X + Y$.

2.17. Suppose X and Y are independent and have a geometric distribution with success probability p. Find the distribution of $X + Y$.

2.18. Suppose $X_1, \ldots, X_n$ are independent and have a geometric distribution with parameter p. Find the distribution of $T = X_1 + \cdots + T_n$.

2.19. Suppose $X =$ uniform on $(0,1)$ and $Y =$ uniform on $(0,2)$ are independent. Find the density function of $X + Y$.

2.20. Suppose X and Y are independent and have density function $2x$ for $x \in (0, 1)$ and 0 otherwise. Find the density function of $X + Y$.

2.21. Suppose X_1, X_2, X_3 are independent and uniform on $(0,1)$. Find the density function of $S = X_1 + X_2 + X_3$. You can save yourself some work by noting that the density will be symmetric about $3/2$.

2.22. Suppose X and Y are independent, X is exponential(λ), and Y is uniform on $(0, 1)$. Find the density function for $X + Y$.

2.23. Suppose X and Y are independent, X has distribution function F, and Y is uniform on $(0, 1)$. (a) Find the density function for $X + Y$. Give easy solutions to the the previous two exercises by considering the special cases in which (b) X is exponential and (c) X has the triangular distribution of Example 2.11.

2.24. Suppose X and Y are independent and have exponential distributions with parameters $\lambda < \mu$. Find the density function of $X + Y$.

3. Expected Value, Moments

If X has a discrete distribution, then the **expected value** of $h(X)$ is

(3.1) $$Eh(X) = \sum_x h(x) P(X = x)$$

When $h(x) = x$ this reduces to EX, the expected value, or **mean of X**, a quantity that is often denoted by μ or sometimes μ_X to emphasize the random variable being considered. When $h(x) = x^k$, $Eh(X) = EX^k$ is the kth **moment**. When $h(x) = (x - EX)^2$,

$$Eh(X) = E(X - EX)^2 = EX^2 - (EX)^2$$

is called the **variance** of X. It is often denoted by $\text{var}(X)$ or σ_X^2. The variance is a measure of how spread out the distribution is. However, if X has the units of feet then the variance has units of feet^2, so the **standard deviation** $\sigma(X) = \sqrt{\text{var}(X)}$, which has again the units of feet, gives a better idea of the "typical" deviation from the mean than the variance does.

Example 3.1. Roll one die. $P(X = x) = 1/6$ for $x = 1, 2, 3, 4, 5, 6$ so

$$EX = (1 + 2 + 3 + 4 + 5 + 6) \cdot \frac{1}{6} = \frac{21}{6} = 3\frac{1}{2}$$

In this case the expected value is just the average of the six possible values.

$$EX^2 = (1^2 + 2^2 + 3^2 + 4^2 + 5^2 + 6^2) \cdot \frac{1}{6} = \frac{91}{6}$$

so the variance is $91/6 - 49/4 = 70/24$. Taking the square root we see that the standard deviation is 1.71. The three possible deviations, in the sense of $|X - EX|$, are 0.5, 1.5, and 2.5 with probability $1/3$ each, so 1.71 is indeed a reasonable approximation for the typical deviation from the mean.

Example 3.2. Poisson distribution. Suppose

$$P(X = k) = e^{-\lambda} \lambda^k / k! \quad \text{for } k = 0, 1, 2, \ldots$$

The $k = 0$ term makes no contribution to the sum, so

$$EX = \sum_{k=1}^{\infty} k e^{-\lambda} \frac{\lambda^k}{k!} = \lambda \sum_{k=1}^{\infty} e^{-\lambda} \frac{\lambda^{k-1}}{(k-1)!} = \lambda$$

The last equality following from the fact that $\sum_{k=1}^{\infty} P(X = (k-1)) = 1$. To compute the variance, we begin by imitating the last calculation to conclude

$$EX(X - 1) = \sum_{k=2}^{\infty} k(k-1) e^{-\lambda} \frac{\lambda^k}{k!} = \lambda^2 \sum_{k=2}^{\infty} e^{-\lambda} \frac{\lambda^{k-2}}{(k-2)!} = \lambda^2$$

From this it follows that $EX^2 = EX(X - 1) + EX = \lambda^2 + \lambda$, and the variance is $EX^2 - (EX)^2 = \lambda$. Notice that the standard deviation is $\sqrt{\lambda}$ which is much smaller than the mean λ when λ is large.

Example 3.3. Geometric distribution. Suppose

$$P(N = k) = p(1-p)^{k-1} \quad \text{for } k = 1, 2, \ldots$$

Starting with the sum of the geometric series

(3.2) $$(1-\theta)^{-1} = \sum_{n=0}^{\infty} \theta^n$$

and then differentiating twice and discarding terms that are 0, gives

$$(1-\theta)^{-2} = \sum_{n=1}^{\infty} n\theta^{n-1} \quad \text{and} \quad 2(1-\theta)^{-3} = \sum_{n=2}^{\infty} n(n-1)\theta^{n-2}$$

Using these with $\theta = 1 - p$, we see that

$$EN = \sum_{n=1}^{\infty} n(1-p)^{n-1} p = p/p^2 = \frac{1}{p}$$

$$EN(N-1) = \sum_{n=2}^{\infty} n(n-1)(1-p)^{n-1} p = 2p^{-3}(1-p)p = \frac{2(1-p)}{p^2}$$

Then reasoning as in the previous example gives

$$\text{var}(N) = EN(N-1) + EN - (EN)^2$$
$$= \frac{2(1-p)}{p^2} + \frac{p}{p^2} - \frac{1}{p^2} = \frac{(1-p)}{p^2} \qquad \square$$

The definition of expected value generalizes in the usual way to continuous random variables. We replace the probability function by the density function and the sum by an integral

(3.3) $$Eh(X) = \int h(x) P(X = x)\, dx$$

Example 3.4. Uniform distribution on (a,b). Suppose X has density function $P(X = x) = 1/(b-a)$ for $a < x < b$ and 0 otherwise. In this case

$$EX = \int_a^b \frac{x}{b-a}\, dx = \frac{b^2 - a^2}{2(b-a)} = \frac{(b+a)}{2}$$

since $b^2 - a^2 = (b-a)(b+a)$. Notice that $(b+a)/2$ is the midpoint of the interval and hence the natural choice for the average value of X. A little more calculus gives

$$EX^2 = \int_a^b \frac{x^2}{b-a}\, dx = \frac{b^3 - a^3}{3(b-a)} = \frac{b^2 + ba + a^2}{3}$$

since $b^3 - a^3 = (b-a)(b^2 + ba + a^2)$. Squaring our formula for EX gives $(EX)^2 = (b^2 + 2ab + a^2)/4$, so

$$\text{var}(X) = (b^2 - 2ab + a^2)/12 = (b-a)^2/12 \qquad \square$$

Example 3.5. Exponential distribution. Suppose X has density function $P(X = x) = \lambda e^{-\lambda x}$ for $x \geq 0$ and 0 otherwise. To compute the expected value, we use the integration by parts formula:

$$(3.4) \qquad \int_a^b g(x) h'(x) \, dx = g(x) h(x) \Big|_a^b - \int_a^b g'(x) h(x) \, dx$$

with $g(x) = x$ and $h'(x) = \lambda e^{-\lambda x}$. Since $g'(x) = 1$, $h(x) = -e^{-\lambda x}$ (3.4) implies

$$EX = \int_0^\infty x \lambda e^{-\lambda x} \, dx$$
$$= -xe^{-\lambda x} \Big|_0^\infty + \int_0^\infty e^{-\lambda x} \, dx = 0 + 1/\lambda$$

where to evaluate the last integral we used $\int_0^\infty \lambda e^{-\lambda x} \, dx = 1$. The second moment is more of the same

$$EX^2 = \int_0^\infty x^2 \lambda e^{-\lambda x} \, dx$$
$$= -x^2 e^{-\lambda x} \Big|_0^\infty + \int_0^\infty 2x e^{-\lambda x} \, dx = 0 + 2/\lambda^2$$

where we have used the result for the mean to evaluate the last integral. The variance is thus

$$\text{var}(X) = EX^2 - (EX)^2 = 1/\lambda^2 \qquad \square$$

To help explain the answers we have found in the last two examples we will now introduce some important properties of expected value.

(3.5) Theorem. *If $X_1, \ldots, X_n$ are any random variables, then*

$$E(X_1 + \cdots + X_n) = EX_1 + \cdots + EX_n$$

(3.6) Theorem. *If $X_1, \ldots, X_n$ are independent, then*

$$\text{var}(X_1 + \cdots + X_n) = \text{var}(X_1) + \cdots + \text{var}(X_n)$$

(3.7) Theorem. *If c is a real number, then*

(a) $E(X + c) = EX + c$ (b) $\text{var}(X + c) = \text{var}(X)$
(c) $E(cX) = cEX$ (d) $\text{var}(cX) = c^2 \text{var}(X)$

Example 3.6. Uniform distribution on (a,b). If X is uniform on the interval $[(a-b)/2, (b-a)/2]$ then $EX = 0$ by symmetry. If $c = (a+b)/2$, then $Y = X + c$ is uniform on $[a, b]$, so it follows from (a) and (b) of (3.7) that

$$EY = EX + c = (a+b)/2 \qquad \text{var}(Y) = \text{var}(X)$$

From the second formula we see that the variance of the uniform distribution will only depend on the length of the interval. To see that it will be a multiple of $(b-a)^2$ note that $Z = X/(b-a)$ is uniform on $[-1/2, 1/2]$ and then use part (d) of (3.7) to conclude $\text{var}(X) = (b-a)^2 \text{var}(Z)$. Of course one needs calculus to conclude that

$$\text{var}(Z) = EZ^2 = \int_{-1/2}^{1/2} x^2\, dx = \left.\frac{x^3}{3}\right|_{-1/2}^{1/2} = \frac{1}{12} \qquad \square$$

Example 3.7. Exponential distribution. A similar scaling explains the results we found in Example 3.5. Changing variables $x = y/\lambda$, $\lambda\, dx = dy$,

$$\int_0^\infty x^k \lambda e^{-\lambda x}\, dx = \frac{1}{\lambda^k} \int_0^\infty y^k e^{-y}\, dy$$

What underlies this relationship between the moments is the fact that if Y has an exponential(1) distribution and $X = Y/\lambda$ then X has an exponential(λ) distribution. Using (3.7) now, it follows that

$$EX = EY/\lambda \quad \text{and} \quad \text{var}(X) = \text{var}(Y)/\lambda^2$$

Again, we have to resort to calculus to show that $EY = 1$ and $EY^2 = 1$ but the scaling relationship tells us the dependence of the answer on λ. $\quad\square$

Using (3.5) and (3.6), we can now derive the mean and variance for two more examples.

Example 3.8. Gamma distribution. Let $t_1, \ldots t_n$ be independent and have the exponential(λ) distribution. By a result in Example 2.12,

$$T_n = t_1 + \cdots + t_n \quad \text{has a gamma}(n, \lambda) \text{ distribution}$$

From (3.5), (3.6), and Example 3.5 it follows that

$$ET_n = nEt_i = n/\lambda \qquad \mathrm{var}(T_n) = n\mathrm{var}(t_i) = n/\lambda^2 \qquad \square$$

Example 3.9. Binomial distribution. If we perform an experiment n times and on each trial there is a probability p of success, then the number of successes S_n has

$$P(S_n = k) = \binom{n}{k} p^k (1-p)^{n-k} \qquad \text{for } k = 0, \ldots, n$$

To compute the mean and variance we begin with the case $n = 1$. Writing X instead of S_1 to simplify notation, we have $P(X = 1) = p$ and $P(X = 0) = 1-p$, so

$$EX = p \cdot 1 + (1-p) \cdot 0 = p$$
$$EX^2 = p \cdot 1^2 + (1-p) \cdot 0^2 = p$$
$$\mathrm{var}(X) = EX^2 - (EX)^2 = p - p^2 = p(1-p)$$

To compute the mean and variance of S_n, we observe that if $X_1, \ldots, X_n$ are independent and have the same distribution as X, then $X_1 + \cdots + X_n$ has the same distribution as S_n. Intuitively, this holds since $X_i = 1$ means one success on the ith trial so the sum counts the total number of success. Using (3.5) and (3.6), we have

$$ES_n = nEX = np \qquad \mathrm{var}(S_n) = n\mathrm{var}(X) = np(1-p) \qquad \square$$

Our final topic is the **moment-generating function** defined by

(3.8) $$\varphi_X(t) = Ee^{tX}$$

The function $\varphi_X(t)$ gets its name from the fact that it generates moments. Differentiating gives $\varphi'_X(t) = E(Xe^{tX})$, $\varphi''_X(t) = E(X^2 e^{tX})$, etc. So using $\varphi^{(k)}(t)$ to denote the kth derivative.

(3.9) $$\varphi_X^{(k)}(t) = E(X^k e^{tX}) \quad \text{and} \quad \varphi_X^{(k)}(0) = E(X^k)$$

The main reason for our interest comes from the following fact:

(3.10) **Theorem.** *If X and Y are independent then*

$$\varphi_{X+Y}(t) = \varphi_X(t)\varphi_Y(t)$$

Proof. Calculus tells that

(3.11) $$e^{a+b} = e^a \cdot e^b$$

Using this with $a = tX$, $b = tY$, noting that e^{tX} and e^{tY} are independent by (2.13) and then using (2.12) we have

$$\varphi_{X+Y}(t) = E(e^{t(X+Y)}) = E(e^{tX}e^{tY})$$
$$= Ee^{tX}Ee^{tY} = \varphi_X(t)\varphi_Y(t)$$

To see the usefulness of (3.10) we consider:

Example 3.10. Poisson distribution. We have $P(X = k) = e^{-\lambda}\lambda^k/k!$ for $k = 0, 1, 2, \ldots$, so

$$\varphi_X(t) = \sum_{k=0}^{\infty} e^{-\lambda}\frac{\lambda^k}{k!}e^{tk} = e^{-\lambda}\sum_{k=0}^{\infty}\frac{(\lambda e^t)^k}{k!} = \exp(-\lambda(1-e^t))$$

From this we see that if $X = \text{Poisson}(\lambda)$ and $Y = \text{Poisson}(\mu)$ are independent

$$\varphi_{X+Y}(t) = \exp(-\lambda(1-e^t)) \cdot \exp(-\mu(1-e^t)) = \exp(-(\lambda+\mu)(1-e^t))$$

so $X + Y = \text{Poisson}(\lambda + \mu)$. □

Technical detail. In the last step we have taken for granted the fact that the moment generating function determines the distribution. If, for example,

$$P(X = x) = \frac{(\rho-1)}{2}(1+|x|)^{-\rho}$$

where $\rho > 1$, then $\varphi_X(t) = \infty$ for $t \neq 0$ and $\varphi_X(0) = 1$, so we cannot distinguish these distributions by looking at their completely uninformative moment generating function. However,

(3.12) If $\varphi_X(t)$ is finite for $|t| < \epsilon$ for some $\epsilon > 0$, then knowing φ_X determines the distribution.

This condition holds for all the examples we will consider.

The moment generating function can be defined for any random variable X. When X takes only nonnegative integer values, it is sometimes more convenient to use the **generating function**:

$$\gamma_X(z) = E(z^X) = \sum_{k=0}^{\infty} z^k P(X = k)$$

Since $P(Z = k) \geq 0$ and sums to one, $\gamma_X(z)$ will always make sense for $0 \leq z \leq 1$. Otherwise this is just a change of variables from the moment generating function:

(3.13) $$\varphi_X(t) = \gamma_X(e^t) \qquad \gamma_X(z) = \varphi_X(\ln z)$$

so it has similar properties with slightly different details. For example, differentiating the generating function gives

$$\gamma'_X(z) = \sum_{k=1}^{\infty} k z^{k-1} P(Z = k)$$

$$\gamma''_X(z) = \sum_{k=2}^{\infty} k(k-1) z^{k-2} P(Z = k)$$

$$\gamma_X^{(j)}(z) = \sum_{k=j}^{\infty} k(k-1) \cdots (k-j+1) z^{k-j} P(Z = k)$$

so setting $z = 1$,

(3.14) $$\gamma_X^{(j)}(1) = E X(X-1) \cdots (X-j+1)$$

In the case of the Poisson, using (3.13) and the result of Example 3.10 gives

$$\gamma_X(z) = \exp(-\lambda(1-z))$$

Differentiating, $\gamma'_X(z) = \lambda \exp(-\lambda(1-z))$ and $\gamma''_X(z) = \lambda^2 \exp(-\lambda(1-z))$, so

$$\gamma_X^{(k)}(z) = \lambda^k \exp(-\lambda(1-z))$$
$$E X(X-1) \cdots (X-k+1) = \gamma_X^{(k)}(1) = \lambda^k$$

EXERCISES

3.1. You want to invent a gambling game in which a person rolls two dice and is paid some money if the sum is 7, but otherwise he loses his money. How much should you pay him for winning a $1 bet if you want this to be a fair game, that is, to have expected value 0?

3.2. In the game of Keno, the casino chooses 20 numbers out of $\{1, \ldots, 80\}$. One possible bet you can make in this game is to choose 3 of the 80 numbers and bet $0.60. You win $ 26 if all three numbers are chosen by the casino, $ 0.60 if two are, and $0 otherwise. Compute the expected value of your winnings.

3.3. Every Sunday afternoon two children, Al and Betty, take turns trying to fly their kites. Suppose Al is successful with probability 1/5 and Betty is successful with probability 1/3. How many Sundays on the average do they have to wait until at least one kite flies?

3.4. Suppose we roll a die repeatedly and let T_k be the number of rolls until some number appears k times in a row. (a) Find a formula that relates ET_k to ET_{k-1}. (b) Compute ET_2 and ET_3.

3.5. Suppose X has density function $3x^2$ for $0 < x < 1$, 0 otherwise. Find the mean, variance, and standard deviation of X.

3.6. Suppose X has density function $3x^{-4}$ for $x > 1$, 0 otherwise. Find the mean, variance, and standard deviation of X.

3.7. Suppose X has density function $cx(2-x)$ for $0 < x < 2$, and 0 otherwise. (a) Pick c to make this a probability density. (b) Find the mean, variance, and standard deviation of X.

3.8. Suppose X has density function $(\lambda/2)e^{-\lambda|x|}$. Find the mean and variance of X.

3.9. Show that $E(X-c)^2$ is minimized by taking $c = EX$.

3.10. Can we have a random variable with $EX = 3$ and $EX^2 = 8$?

3.11. Suppose X and Y are independent with $\text{var}(X) = 2$ and $\text{var}(Y) = 1$ finite. Find $\text{var}(3X - 4Y + 5)$.

3.12. In a class with 18 boys and 12 girls, boys have probability 1/3 of knowing the answer and girls have probability 1/2 of knowing the answer to a typical question the teacher asks. Assuming that whether or not the students know the answer are independent events, find the mean and variance of the number of students who know the answer.

3.13. Suppose we draw 13 cards out of a deck of 52. What is the expected value of the number of aces we get?

3.14. Suppose we pick 3 students at random from a class with 10 boys and 15 girls. Let X be the number of boys selected and Y be the number of girls selected. Find $E(X - Y)$.

3.15. Let N_k be the number of independent trials we need to get k successes when success has probability p. Find EN_k.

3.16. Suppose we put m balls randomly into n boxes. What is the expected number of empty boxes?

3.17. Suppose we put 7 balls randomly into 5 boxes. What is the expected number of empty boxes?

3.18. Suppose Noah started with n pairs of animals on the ark and m of them died. If we suppose that fate chose the m animals at random, what is the expected number of complete pairs that are left?

3.19. Suppose Noah started with 50 pairs of animals on the ark and 20 of them died. If we suppose that fate chose the 20 animals at random, what is the expected number of complete pairs that are left?

3.20. Suppose $X =$ Poisson(λ). Find $E(1/(X+1))$.

3.21. Suppose we roll two dice and let X and Y be the numbers they show. Find the expected value of (a) $X + Y$, (b) $\max\{X,Y\}$, (c) $\min\{X,Y\}$, (d) $|X - Y|$.

3.22. Suppose X and Y are independent and exponential(λ). Find $E|X - Y|$.

3.23. Suppose U and V are independent and uniform on (0,1). Find EU^2 and $E(U - V)^2$.

3.24. Suppose X has the standard normal distribution. Use integration by parts to show that $EX^k = (k-1)EX^{k-2}$ and then conclude that for all integers n, $EX^{2n-1} = 0$ and $EX^{2n} = (2n-1)(2n-3)\cdots 3\cdot 1$.

3.25. Suppose X has the standard normal distribution. Find Ee^{tX}. Differentiate to find the mean and variance of X.

3.26. Suppose that X is exponential(λ). (a) Find Ee^{tX}. (b) Differentiate to find the mean and variance of X. (c) Suppose $Y =$ gamma(n, λ). Use (a) to find Ee^{tY}.

3.27. Let X have the standard normal distribution given in Example 2.6 and let $Y = \sigma X + \mu$. (a) Find EY and var(Y). (b) Show that $\varphi_Y(t) = e^{\mu t}\varphi_X(\sigma t)$. (c) Use (b) with the previous exercise to evaluate Ee^{tY}.

3.28. Suppose that X is 1 or 0 with probabilities p and $1 - p$. (a) Find Ee^{tX}. (b) Differentiate to find the mean and variance of X. (c) Suppose $Y =$ binomial(n, λ) Use (a) to find Ee^{tY}.

3.29. Suppose X has a shifted geometric distribution, i.e., $P(X = k) = p(1-p)^k$ for $k = 0,1,2,\ldots$ (a) Find the generating function $\gamma_X(z) = Ez^X$. (b) Differentiate to find the mean and variance of X. (c) Check the results of (b) by comparing with those found in Example 3.3.

1 Markov Chains

1.1. Definitions and Examples

The importance of Markov chains comes from two facts: (i) there are a large number of physical, biological, economic, and social phenomena that can be described in this way, and (ii) there is a well-developed theory that allows us to do computations. We begin with a famous example, then describe the property that is the defining feature of Markov chains.

Example 1.1. Gambler's ruin. Consider a gambling game in which on any turn you win \$1 with probability $p = 0.4$ or lose \$1 with probability $1 - p = 0.6$. Suppose further that you adopt the rule that you quit playing if your fortune reaches \$N. Of course, if your fortune reaches \$0 the casino makes you stop.

Let X_n be the amount of money you have after n plays. I claim that your fortune, X_n has the "Markov property." In words, this means that given the current state, any other information about the past is irrelevant for predicting the next state X_{n+1}. To check this for the gambler's ruin chain, we note that if you are still playing at time n, i.e., your fortune $X_n = i$ with $0 < i < N$, then for any possible history of your wealth $i_{n-1}, i_{n-2}, \ldots i_1, i_0$

$$P(X_{n+1} = i + 1 | X_n = i, X_{n-1} = i_{n-1}, \ldots X_0 = i_0) = 0.4$$

since to increase your wealth by one unit you have to win your next bet. Here we have used $P(B|A)$ for the conditional probability of the event B given that A occurs. Recall that this is defined by

$$P(B|A) = \frac{P(B \cap A)}{P(A)}$$

Turning now to the formal definition, we say that X_n is a discrete time **Markov chain** with **transition matrix** $p(i, j)$ if for any $j, i, i_{n-1}, \ldots i_0$

(1.1) $\qquad P(X_{n+1} = j | X_n = i, X_{n-1} = i_{n-1}, \ldots, X_0 = i_0) = p(i, j)$

Equation (1.1), also called the "Markov property" says that the conditional probability $X_{n+1} = j$ given the entire history $X_n = i, X_{n-1} = i_{n-1}, \ldots X_1 = i_1, X_0 = i_0$ is the same as the conditional probability $X_{n+1} = j$ given only the previous state $X_n = i$. This is what we mean when we say that "any other information about the past is irrelevant for predicting X_{n+1}."

In formulating (1.1) we have restricted our attention to the **temporally homogeneous** case in which the **transition probability**

$$p(i,j) = P(X_{n+1} = j | X_n = i)$$

does not depend on the time n. Intuitively, the transition probability gives the rules of the game. It is the basic information needed to describe a Markov chain. In the case of the gambler's ruin chain, the transition probability has

$$p(i, i+1) = 0.4, \quad p(i, i-1) = 0.6, \quad \text{if } 0 < i < N$$
$$p(0,0) = 1 \qquad p(N,N) = 1$$

When $N = 5$ the matrix is

	0	1	2	3	4	5
0	1.0	0	0	0	0	0
1	0.6	0	0.4	0	0	0
2	0	0.6	0	0.4	0	0
3	0	0	0.6	0	0.4	0
4	0	0	0	0.6	0	0.4
5	0	0	0	0	0	1.0

or the chain by be represented pictorially as

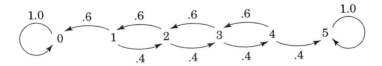

Example 1.2. Ehrenfest chain. This chain originated in physics as a model for two cubical volumes of air connected by a small hole. In the mathematical version, we have two "urns," i.e., two of the exalted trash cans of probability theory, in which there are a total of N balls. We pick one of the N balls at random and move it to the other urn.

Let X_n be the number of balls in the "left" urn after the nth draw. It should be clear that X_n has the Markov property; i.e., if we want to guess the state at time $n+1$, then the current number of balls in the left urn X_n, is the only relevant information from the observed sequence of states $X_n, X_{n-1}, \ldots X_1, X_0$. To check this we note that

$$P(X_{n+1} = i+1 | X_n = i, X_{n-1} = i_{n-1}, \ldots X_0 = i_0) = (N-i)/N$$

since to increase the number we have to pick one of the $N-i$ balls in the other urn. The number can also decrease by 1 with probability i/N. In symbols, we have computed that the transition probability is given by

$$p(i, i+1) = (N-i)/N, \quad p(i, i-1) = i/N \quad \text{for } 0 \le i \le N$$

with $p(i,j) = 0$ otherwise. When $N=5$, for example, the matrix is

	0	1	2	3	4	5
0	0	5/5	0	0	0	0
1	1/5	0	4/5	0	0	0
2	0	2/5	0	3/5	0	0
3	0	0	3/5	0	2/5	0
4	0	0	0	4/5	0	1/5
5	0	0	0	0	5/5	0

Here we have written 1 as 5/5 to emphasize the pattern in the diagonals of the matrix.

In most cases Markov chains are described by giving their transition probabilities.

Example 1.3. Weather chain. Let X_n be the weather on day n in Ithaca, NY. Since this course is taught in the "spring" semester, we will use three states: $1 = snowy$, $2 = cloudy$, $3 = sunny$. The weather is certainly not a Markov chain. For example, if the previous two days were sunny, then it is more likely that there is a special atmospheric condition like a high-pressure area located in the region, which favors sunshine, so we would have a higher probability of a third sunny day.

Even though the weather is not exactly a Markov chain, we can propose a Markov chain model for the weather by writing down a guess for the transition probability

	1	2	3
1	.4	.6	0
2	.2	.5	.3
3	.1	.7	.2

The table says, for example, the probability a snowy day (state 1) is followed by a cloudy day (state 2) is $p(1,2) = 0.6$. There is nothing very special about these probabilities. You can write down any 3×3 matrix, provided that the entries satisfy:

(i) $p(i,j) \geq 0$, since they are probabilities.

(ii) $\sum_j p(i,j) = 1$, since when $X_n = i$, X_{n+1} will be in some state j.

The equation in (ii) is read "sum $p(i,j)$ over all possible values of j." In words the last two conditions say: the entries of the matrix are nonnegative and each ROW of the matrix sums to 1.

Any matrix with properties (i) and (ii) gives rise to a Markov chain, X_n. To construct the chain we can think of playing a board game. When we are in state i, we roll a die (or generate a random number on a computer) to pick the next state, going to j with probability $p(i,j)$.

The next four examples will help explain why we want to study Markov chains. At the moment we don't have any theory to work with, so all we can do is to write down the transition probability and then ask a question that we will answer later.

Example 1.4. Repair chain. A machine has three critical parts that are subject to failure, but can function as long as two of these parts are working. When two are broken, they are replaced and the machine is back to working order the next day. To formulate a Markov chain model we declare its state space to be the parts that are broken $\{0, 1, 2, 3, 12, 13, 23\}$. If we assume that parts 1, 2, and 3 fail with probabilities .01, .02, and .04, but no two parts fail on the same day, then we arrive at the following transition matrix:

	0	1	2	3	12	13	23
0	.93	.01	.02	.04	0	0	0
1	0	.94	0	0	.02	.04	0
2	0	0	.95	0	.01	0	.04
3	0	0	0	.97	0	.01	.02
12	1	0	0	0	0	0	0
13	1	0	0	0	0	0	0
23	1	0	0	0	0	0	0

If we own a machine like this, then it is natural to ask about the long-run cost per day to operate it. For example, we might ask:

Q. If we are going to operate the machine for 1800 days (about 5 years), then how many parts of types 1, 2, and 3 will we use?

Chapter 1 Markov Chains

Example 1.5. Inventory chain. An electronics store sells a video game system. If at the end of the day, the number of units they have on hand is 1 or 0, they order enough new units so their total on hand is 5. For simplicity we assume that the new merchandise arrives before the store opens the next day. Let X_n be the number of units on hand at the end of the nth day. If we assume that the number of customers who want to buy a video game system each day is 0, 1, 2, or 3 with probabilities .3, .4, .2, and .1, then we have the following transition matrix:

	0	1	2	3	4	5
0	0	0	.1	.2	.4	.3
1	0	0	.1	.2	.4	.3
2	0.2	.4	.3	0	0	0
3	.1	.2	.4	.3	0	0
4	0	.1	.2	.4	.3	0
5	0	0	.1	.2	.4	.3

The original value 3 is correct since the row elements must sum up to 1.

To explain the entries we note that if $X_n = 0$ or 1, we order enough to have 5 units at the beginning of the next day. If $X_n = 2$ and the demand is 3 or more, or if $X_n = 3$ and the demand is 4, we end up with 0 units at the end of the day and at least one unhappy customer.

This chain is an example of an s, S inventory control policy with $s = 1$ and $S = 5$. That is, when the stock on hand falls to s or below we order enough to bring it back up to S. Let D_{n+1} be the demand on day $n + 1$. Introducing notation for the **positive part** of a real number,

$$x^+ = \max\{x, 0\} = \begin{cases} x & \text{if } x > 0 \\ 0 & \text{if } x \leq 0 \end{cases}$$

then we can write the chain in general as

$$X_{n+1} = \begin{cases} (X_n - D_{n+1})^+ & \text{if } X_n > s \\ (S - D_{n+1})^+ & \text{if } X_n \leq s \end{cases}$$

In this context we might be interested in:

Q. Suppose we make $12 profit on each unit sold but it costs $2 a day to store items. What is the long-run profit per day of this inventory policy? How do we choose s and S to maximize profit?

Example 1.6. Branching processes. Consider a population in which each individual in the nth generation independently gives birth, producing k children (who are members of generation $n + 1$) with probability p_k. The number of individuals in generation n, X_n, can be any nonnegative integer, so the state space is $\{0, 1, 2, \ldots\}$. If we let $Y_1, Y_2, \ldots$ be independent random variables with $P(Y_m = k) = p_k$, then we can write the transition probability as

$$p(i, j) = P(Y_1 + \cdots + Y_i = j) \quad \text{for } i > 0 \text{ and } j \geq 0$$

When there are no living members of the population, no new ones can be born, so $p(0, 0) = 1$. Here 0 is an **absorbing state**. Once the process hits 0, it can never leave. Thus it is natural to ask:

Q. What is the probability that the species does not die out, i.e., the process avoids absorption at 0?

A slight variant of this question provided the original motivation of Galton and Watson. They were interested in the probability a family name would die out, so they formulated a branching process model in which p_k was the probability a man would have k male children.

Example 1.7. Wright–Fisher model. Thinking of a population of $N/2$ diploid individuals who have two copies of each of their chromosomes, or of N haploid individuals who have one copy, we consider a fixed population of N genes that can be one of two types: A or a. In the simplest version of this model the population at time $n+1$ is obtained by drawing with replacement from the population at time n. In this case if we let X_n be the number of A alleles at time n, then X_n is a Markov chain with transition probability

$$p(i, j) = \binom{N}{j} \left(\frac{i}{N}\right)^j \left(1 - \frac{i}{N}\right)^{N-j}$$

since the right-hand side is the binomial distribution for N independent trials with success probability i/N.

In this model the states 0 and N that correspond to fixation of the population in the all a or all A states are absorbing states, so it is natural to ask:

Q1. Starting from i of the A alleles and $N - i$ of the a alleles, what is the probability that the population fixates in the all A state?

To make this simple model more realistic we can introduce the possibility of mutations: an A that is drawn ends up being an a in the next generation with probability u, while an a that is drawn ends up being an A in the next generation with probability v. In this case the probability an A is produced by a given draw is

$$\rho_i = \frac{i}{N}(1 - u) + \frac{N - i}{N} v$$

but the transition probability still has the binomial form

$$p(i,j) = \binom{N}{j}(\rho_i)^j (1-\rho_i)^{N-j}$$

If u and v are both positive, then 0 and N are no longer absorbing states, so we ask:

Q2. Does the genetic composition settle down to an equilibrium distribution as time $t \to \infty$?

To explain what it might mean for the Markov chain to "settle down to an equilibrium distribution," we consider a simpler example with only three states.

Example 1.8. Income classes. Suppose that from one generation to the next families change their income group Low, Middle, or High according to the following Markov chain.

	L	M	H
L	0.6	0.3	0.1
M	0.2	0.7	0.1
H	0.1	0.3	0.6

It is natural to ask:

Q1. Do the fractions of the population in the three income classes stabilize as time goes on?

As the reader might guess the answer to this question is "Yes." However, knowing this just leads to another question:

Q2. How can we compute the limit proportions from the transition matrix?

1.2. Multistep Transition Probabilities

The transition probability $p(i,j) = P(X_{n+1} = j | X_n = i)$ gives the probability of going from i to j in one step. Our goal in this section is to compute the probability of going from i to j in $m > 1$ steps:

$$p^m(i,j) = P(X_{n+m} = j | X_n = i)$$

To warm up we recall the transition probability of the weather chain:

	1	2	3
1	.4	.6	0
2	.2	.5	.3
3	.1	.7	.2

Section 1.2 Multistep Transition Probabilities

and consider the following concrete question:

Problem 2.1. Today is Tuesday and it is cloudy (state 2). What is the probability Wednesday is sunny (state 3) and Thursday is snowy (state 1)?

Solution. Intuitively, the Markov property implies that starting from state 2 the probability of jumping to 3 and then to 1 is given by $p(2,3)p(3,1)$. To get this conclusion from the definitions, we note that using the definition of conditional probability,

$$P(X_2 = 1, X_1 = 3 | X_0 = 2) = \frac{P(X_2 = 1, X_1 = 3, X_0 = 2)}{P(X_0 = 2)}$$
$$= \frac{P(X_2 = 1, X_1 = 3, X_0 = 2)}{P(X_1 = 3, X_0 = 2)} \cdot \frac{P(X_1 = 3, X_0 = 2)}{P(X_0 = 2)}$$
$$= P(X_2 = 1 | X_1 = 3, X_0 = 2) \cdot P(X_1 = 3 | X_0 = 2)$$

By the Markov property (1.1) the last expression is

$$P(X_2 = 1 | X_1 = 3) \cdot P(X_1 = 3 | X_0 = 2) = p(2,3)p(3,1)$$

Moving on to the real question:

Problem 2.2. What is the probability Thursday is snowy (1) given that Tuesday is cloudy (2)?

Solution. To do this we simply have to consider the three possible states for Wednesday and use the solution of the previous problem.

$$P(X_2 = 1 | X_0 = 2) = \sum_{k=1}^{3} P(X_2 = 1, X_1 = k | X_0 = 2) = \sum_{k=1}^{3} p(2,k)p(k,1)$$
$$= (.2)(.4) + (.5)(.2) + (.3)(.1) = .08 + .10 + .03 = .21$$

There is nothing special here about the states 2 and 1 here. By the same reasoning,

$$P(X_2 = j | X_0 = i) = \sum_{k=1}^{3} p(i,k) \, p(k,j)$$

The right-hand side of the last equation gives the (i,j)th entry of the matrix p is multiplied by itself, so the two step transition probability p^2 is simply the second power of the transition matrix p. From this you can probably leap to the next conclusion:

Chapter 1 Markov Chains

(2.1) Theorem. *The m step transition probability $P(X_{n+m} = j | X_n = i)$ is the mth power of the transition matrix p.*

The key ingredient in proving this is the:

(2.2) Chapman–Kolmogorov equation

$$p^{m+n}(i,j) = \sum_k p^m(i,k)\, p^n(k,j)$$

Once this is proved, (2.1) follows, since taking $n = 1$ in (2.2), we see that

$$p^{m+1}(i,j) = \sum_k p^m(i,k)\, p(k,j)$$

That is, the $m+1$ step transition probability is the m step transition probability times p. When $m = 1$ this says again that p^2 is p times p, but starting with $m = 2$ this provides useful information: p^3 is p^2 times p and hence the third power of p. Repeating, we conclude that (2.1) is true for all $m \geq 4$.

Why is (2.2) true? To go from i to j in $m+n$ steps, we have to go from i to some state k in m steps and then from k to j in n steps. The Markov property implies that the two parts of our journey are independent. □

Of course, the *independence* here is the mysterious part. To show without a doubt that this is true we give the

Proof of (2.2). We do this by combining the solutions of Problems 2.1 and 2.2. Breaking things down according to the state at time m,

$$P(X_{m+n} = j | X_0 = i) = \sum_k P(X_{m+n} = j, X_m = k | X_0 = i)$$

Using the definition of conditional probability as in the solution of Problem 2.1,

$$P(X_{m+n} = j, X_m = k | X_0 = i) = \frac{P(X_{m+n} = j, X_m = k, X_0 = i)}{P(X_0 = i)}$$

$$= \frac{P(X_{m+n} = j, X_m = k, X_0 = i)}{P(X_m = k, X_0 = i)} \cdot \frac{P(X_m = k, X_0 = i)}{P(X_0 = i)}$$

$$= P(X_{m+n} = j | X_m = k, X_0 = i) \cdot P(X_m = k | X_0 = i)$$

By the Markov property (1.1) the last expression is

$$= P(X_{m+n} = j | X_m = k) \cdot P(X_m = k | X_0 = i) = p^m(i,k) p^n(k,j)$$

and we have proved (2.2). □

Having established (2.2), we now return to computations. We begin computing the entire two-step transition probability for the

Example 2.1. Weather chain. To compute, for example, $p^2(1,3)$, the probability of going from 1 to 3 in two steps we need to compute the entry in the first row and third column of $p \times p$. In symbols,

$$p^2(1,3) = \sum_k p(1,k)\, p(k,3)$$

In words, we multiply the first row by the third column of p. In matrix notation:

$$\begin{pmatrix} .4 & .6 & 0 \\ . & . & . \\ . & . & . \end{pmatrix} \begin{pmatrix} . & . & 0 \\ . & . & .3 \\ . & . & .2 \end{pmatrix} = \begin{pmatrix} . & . & .18 \\ . & . & . \\ . & . & . \end{pmatrix}$$

since $(.4)(0) + (.6)(.3) + (0)(.2) = 0.18$. Repeating this computation for the other matrix elements we have:

$$\begin{pmatrix} .4 & .6 & 0 \\ .2 & .5 & .3 \\ .1 & .7 & .2 \end{pmatrix} \cdot \begin{pmatrix} .4 & .6 & 0 \\ .2 & .5 & .3 \\ .1 & .7 & .2 \end{pmatrix} = \begin{pmatrix} .28 & .54 & .18 \\ .21 & .58 & .21 \\ .20 & .55 & .25 \end{pmatrix}$$

To check the rest of the first row, for example, we note

$$(.4)(.4) + (.6)(.2) + (0)(.1) = .28$$
$$(.4)(.6) + (.6)(.5) + (0)(.7) = .54$$

To anticipate Section 1.4, where we will discuss the limiting behavior of p^n as $n \to \infty$, we note that the rows of the matrix are already very similar, while computing higher powers using the Chapman–Kolmogorov equation (2.2) gives

$$p^4 = p^2 \cdot p^2 = \begin{pmatrix} .2278 & .5634 & .2088 \\ .2226 & .5653 & .2121 \\ .2215 & .5645 & .2140 \end{pmatrix}$$

$$p^8 = p^4 \cdot p^4 = \begin{pmatrix} .22355 & .56470 & .21175 \\ .22352 & .56471 & .21177 \\ .22352 & .56471 & .21177 \end{pmatrix}$$

Knowing the answers that will come later, we would like to note that here p^n is converging rapidly to a matrix in which each row is the same. In Example 4.7 we will see that in the limit each row is

$$(19/85, 48/85, 18/85) = (.22353, .56471, .21176)$$

Example 2.2. Gambler's ruin. Suppose for simplicity that $N = 4$ in Example 1.1, so that the transition probability is

$$\begin{array}{c|ccccc} & 0 & 1 & 2 & 3 & 4 \\ \hline 0 & 1.0 & 0 & 0 & 0 & 0 \\ 1 & 0.6 & 0 & 0.4 & 0 & 0 \\ 2 & 0 & 0.6 & 0 & 0.4 & 0 \\ 3 & 0 & 0 & 0.6 & 0 & 0.4 \\ 4 & 0 & 0 & 0 & 0 & 1.0 \end{array}$$

To compute p^2 one row at a time we note:

$p^2(0,0) = 1$ and $p^2(4,4) = 1$, since these are absorbing states.

$p^2(1,3) = (.4)^2 = 0.16$, since the chain has to go up twice.

$p^2(1,1) = (.4)(.6) = 0.24$. The chain must go from 1 to 2 to 1.

$p^2(1,0) = 0.6$. To be at 0 at time 2, the first jump must be to 0.

Leaving the cases $i = 2, 3$ to the reader, we have

$$p^2 = \begin{pmatrix} 1.0 & 0 & 0 & 0 & 0 \\ .6 & .24 & 0 & .16 & 0 \\ .36 & 0 & .48 & 0 & .16 \\ 0 & .36 & 0 & .24 & .4 \\ 0 & 0 & 0 & 0 & 1 \end{pmatrix}$$

Again, to anticipate the developments in the next section we note that with the help of a small computer program one can easily compute.

$$p^{20} = \begin{pmatrix} 1.0 & 0 & 0 & 0 & 0 \\ .87655 & .00032 & 0 & .00022 & .12291 \\ .69186 & 0 & .00065 & 0 & .30749 \\ .41842 & .00049 & 0 & .00032 & .58437 \\ 0 & 0 & 0 & 0 & 1 \end{pmatrix}$$

Looking at the middle row of this matrix we see that when the chain starts at 2, then with high probability the chain is either at 0 or 4. Note also that it is impossible to go from 2 to 3 in 20 steps since the gambler's fortune alternates between being an even and an odd number until it gets stuck at one of the endpoints. This phenomenon is called periodicity and will be addressed in Section 1.4.

The last topic to be considered in this section is:

What happens when the initial state is random? Breaking things down according to the value of the initial state and using the definition of conditional probability

$$P(X_n = y) = \sum_x P(X_0 = x, X_n = y)$$
$$= \sum_x P(X_0 = x) P(X_n = y | X_0 = x)$$

If we introduce $\mu(x) = P(X_0 = x)$, then the last equation can be written as

(2.3) $$P(X_n = y) = \sum_x \mu(x) p^n(x,y)$$

In words, we multiply the transition matrix on the left by the vector μ of initial probabilities. If there are k states, then $p^n(x,y)$ is a $k \times k$ matrix. So to make the matrix multiplication work out right we should take μ as a $1 \times k$ matrix or a "row vector."

For a concrete example consider

Example 2.3. Weather chain. Suppose that the initial distribution: $\mu(1) = .5$, $\mu(2) = .2$, and $\mu(3) = .3$. Multiplying the vector μ times the transition probability gives the vector of probabilities at time 1.

$$(.5 \quad .2 \quad .3) \begin{pmatrix} .4 & .6 & 0 \\ .2 & .5 & .3 \\ .1 & .7 & .2 \end{pmatrix} = (.27 \quad .61 \quad .12)$$

To check the arithmetic note that the first entry is

$$.5(.4) + .2(.2) + .3(.1) = .20 + .04 + .03 = .27$$

1.3. Classification of States

Let $T_y = \min\{n \geq 1 : X_n = y\}$ be the time of the **first return** to y (i.e., being there at time 0 doesn't count), and let

$$g_y = P(T_y < \infty | X_0 = y)$$

be the probability X_n returns to y when it starts at y. Intuitively, the Markov property implies that the probability X_n will return at least twice to y is g_y^2, since after the first return, the chain is at y, the game starts again, and the probability of a second return following the first is again g_y.

To show that the reasoning in the last paragraph is valid, we have to introduce a definition and state a theorem.

We say that T is a **stopping time** if the occurrence (or nonoccurrence) of the event "we stop at time n," $\{T = n\}$ can be determined by looking at the values of the process up to that time: $X_0, \ldots, X_n$.

To see that T_y is a stopping time note that
$$\{T_y = n\} = \{X_1 \neq y, \ldots, X_{n-1} \neq y, X_n = y\}$$
and that the right-hand side can be determined from $X_0, \ldots, X_n$.

Since stopping at time n depends only on the values $X_0, \ldots, X_n$, and in a Markov chain the distribution of the future only depends on the past through the current state, it should not be hard to believe that the Markov property holds at stopping times. This fact can be stated formally as:

(3.1) Strong Markov property. *Suppose T is a stopping time. Given that $T = n$ and $X_T = y$, any other information about $X_0, \ldots X_T$ is irrelevant for predicting the future, and X_{T+k}, $k \geq 0$ behaves like the Markov chain with initial state y.*

Why is this true? To keep things as simple as possible we will show only that
$$P(X_{T+1} = z | X_T = y, T = n) = p(y, z)$$
Let V_n be the set of vectors $(x_0, \ldots, x_n)$ so that if $X_0 = x_0, \ldots, X_n = x_n$, then $T = n$ and $X_T = y$. Breaking things down according to the values of $X_0, \ldots, X_n$ gives
$$P(X_{T+1} = z, X_T = y, T = n) = \sum_{x \in V_n} P(X_{n+1} = z, X_n = x_n, \ldots, X_0 = x_0)$$
$$= \sum_{x \in V_n} P(X_{n+1} = z | X_n = x_n, \ldots, X_0 = x_0) P(X_n = x_n, \ldots, X_0 = x_0)$$
where in the second step we have used the definition of conditional probability $P(B|A) = P(B \cap A)/P(A)$. For any $(x_0, \ldots, x_n) \in A$ we have $T = n$ and $X_T = y$ so $x_n = y$. Using the Markov property, (1.1), and recalling the definition of V_n shows the above
$$P(X_{T+1} = z, T = n, X_T = y) = p(y, z) \sum_{x \in V_n} P(X_n = x_n, \ldots, X_0 = x_0)$$
$$= p(y, z) P(T = n, X_T = y)$$

Section 1.3 Classification of States

Dividing both sides by $P(T = n, X_T = y)$ gives the desired result. □

The strong Markov property implies that the conditional probability we will return one more time given that we have returned $n-1$ times is g_y. This implies that the probability of returning twice is g_y^2, the probability of returning three times is $g_y^2 \cdot g_y = g_y^3$, or in general that the probability of returning n times is g_y^n. At this point, there are two possibilities:

(i) $g_y < 1$: The probability of returning n times is $g_y^n \to 0$ as $n \to \infty$. Thus, eventually the Markov chain does not find its way back to y. In this case the state y is called **transient**, after some point it never reappears in the Markov chain.

(ii) $g_y = 1$: The probability of returning n times $g_y^n = 1$, so the chain returns to y infinitely many times. In this case, the state y is called **recurrent**, it continually recurs in the Markov chain.

To understand these notions, we turn to our examples, beginning with one of our favorites.

Example 3.1. Gambler's ruin. Consider for concreteness the case $N = 4$.

	0	1	2	3	4
0	1	0	0	0	0
1	.6	0	.4	0	0
2	0	.6	0	.4	0
3	0	0	.6	0	.4
4	0	0	0	0	1

We will show that eventually the chain gets stuck in either the bankrupt (0) or happy (4) state. In the terms of our recent definitions, we will show that states $0 < y < 4$ are transient, while the states 0 and 4 are recurrent.

It is easy to check that 0 and 4 are recurrent. Since $p(0,0) = 1$, the chain comes back on the next step with probability one, i.e.,

$$P(T_0 = 1 | X_0 = 0) = 1$$

and hence $g_0 = 1$. A similar argument shows that 4 is recurrent. In general if y is an **absorbing state**, i.e., if $p(y,y) = 1$, then y is a very strongly recurrent state – the chain always stays there.

To check the transience of the interior states, 1, 2, 3, we note that starting from 1, if the chain goes to 0, it will never return to 1, so the probability of never returning to 1,

$$P(T_1 = \infty | X_0 = 1) \geq p(1,0) = 0.6 > 0$$

Similarly, starting from 2, the chain can go to 1 and then to 0, so

$$P(T_2 = \infty | X_0 = 2) \geq p(2,1)p(1,0) = 0.36 > 0$$

Finally for starting from 3, we note that the chain can go immediately to 4 and never return with probability 0.4, so

$$P(T_3 = \infty | X_0 = 3) \geq p(3,4) = 0.4 > 0$$

Important notation. Probabilities like the last four in which we are interested in the behavior of the chain for a fixed initial state will come up often so we will introduce the shorthand

$$P_x(A) = P(A | X_0 = x)$$

Later we will have to consider expected values for this probability and we will denote them by E_x.

Generalizing from our experience with the gambler's ruin chain, we come to a general result that will help us identify transient states. We say that x **communicates with** y and write $x \to y$ if there is a positive probability of reaching y starting from x, that is, the probability

$$\rho_{xy} = P_x(T_y < \infty) > 0$$

Note that the last probability includes not only the possibility of jumping from x to y in one step but also going from x to y after visiting several other states in between.

EXERCISE 3.1. If $x \to y$ and $y \to z$, then $x \to z$.

(3.2) **Theorem.** If x communicates with y, but y does not communicate with x, then x is transient.

Why is this true? By logic (3.2) is equivalent to "if x is recurrent and x communicates with y, then y communicates with x." In fact we will show $\rho_{yx} = 1$. Intuitively this holds since if $\rho_{yx} < 1$, then it would be possible to go from x to y and never return to x, contradicting recurrence of x.

Proof. To get a formal contradiction we have to work a little more carefully. Let $K = \min\{k : p^k(x,y) > 0\}$ be the smallest number of steps we can take to get from x to y. Since $p^K(x,y) > 0$ there must be a sequence $y_1, \ldots y_{K-1}$ so that

$$p(x, y_1)p(y_1, y_2) \cdots p(y_{K-1}, y) > 0$$

Since K is minimal all the $y_i \neq y$ (or there would be a shorter path), and we have
$$P_x(T_x = \infty) \geq p(x, y_1)p(y_1, y_2) \cdots p(y_{K-1}, y)(1 - \rho_{yx})$$
x being recurrent implies that the left side is 0, so if $\rho_{yx} < 1$ we would have a contradiction. □

We will see later that (3.2) allows us to to identify all the transient states when the state space is finite. In the next example we will attack the problem of identifying recurrent states.

Example 3.2. Weather chain. Recall that the transition probability is
$$p = \begin{matrix} .4 & .6 & 0 \\ .2 & .5 & .3 \\ .1 & .7 & .2 \end{matrix}$$

In this case we will use a "bare hands" approach to show that all states are recurrent. This clumsiness should help convince the reader that the theory we will develop in this section is useful. To begin we note that no matter where X_n is, there is a probability of at least .1 of hitting 1 on the next step so $P_1(T_1 > n) \leq (.9)^n$. As $n \to \infty$, $(.9)^n \to 0$ so $P_1(T_1 < \infty) = 1$, i.e., we will return to 1 with probability 1.

The last argument applies even more strongly to state 2, where the probability of jumping to 2 on the next step is always at least .5. However, the argument does not work for state 3 since there the probability of going from 1 to 3 in one step is 0. To get around that problem we note that, as we have computed in Example 2.1,
$$p^2 = \begin{matrix} .28 & .54 & .18 \\ .21 & .58 & .21 \\ .20 & .55 & .25 \end{matrix}$$

From this we see that $P(X_{n+2} = 3 | X_n = x) \geq .18$ for $x = 1, 2, 3$ or in words no matter where we are we have a probability of at least .18 of hitting 3 two time units later. By considering where the chain is at times $2, 4, \ldots, 2k$, which gives us k possibilities to hit 3, it follows that the probability we do not hit 3 before time $2k$ is
$$P_3(T_3 > 2k) \leq (.82)^k$$
Again $(.82)^k$ tends to 0, as $k \to \infty$ so state 3 is recurrent.

The last argument generalizes to the give the following useful fact. The proof will explain the somewhat strange name that has become attached to this result.

44 Chapter 1 Markov Chains

(3.3) **Pedestrian lemma.** Suppose $P_x(T_y \leq k) \geq \alpha > 0$ for all x in the state space S. Then
$$P_x(T_y > nk) \leq (1-\alpha)^n$$

Why is this true? Consider a pedestrian who always has a probability of at least α of being killed when crossing the street, then the probability of completing n successful crossings is at most $(1-\alpha)^n$. □

To test the theory we have developed so far, consider

Example 3.3. A Seven-state chain. Consider the transition matrix:

	1	2	3	4	5	6	7
1	.3	0	0	0	.7	0	0
2	.1	.2	.3	.4	0	0	0
3	0	0	.5	.5	0	0	0
4	0	0	0	.5	0	.5	0
5	.6	0	0	0	.4	0	0
6	0	0	0	0	0	.2	.8
7	0	0	0	1	0	0	0

To identify the states that are recurrent and those that are transient, we begin by drawing a graph that will contain an arc from i to j if $p(i,j) > 0$. We do not worry about drawing the self-loops corresponding to states with $p(i,i) > 0$ since such transitions cannot help the chain get somewhere new.

In the case under consideration we draw arcs from $1 \to 5$, $2 \to 1$, $2 \to 3$, $2 \to 4$, $3 \to 4$, $3 \to 5$, $4 \to 6$, $4 \to 7$, $5 \to 1$, $6 \to 4$, $6 \to 7$, $7 \to 4$.

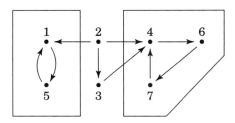

The state 2 communicates with 1, which does not communicate with it, so (3.2) implies that 2 is transient. Likewise 3 communicates with 4, which doesn't communicate with it, so 3 is transient. The remaining states separate into two mutually communicating sets $\{1,5\}$, and $\{4,6,7\}$. To conclude that all these states are recurrent we will introduce two definitions and a fact.

A set A is **closed** if it is impossible to get out, i.e., if $i \in A$ and $j \notin A$ then $p(i,j) = 0$. In Example 3.3, $\{1,5\}$ and $\{4,6,7\}$ are closed sets. Of course their union, $\{1,5,4,6,7\}$ is also closed, as is the whole state space $\{1,2,3,4,5,6,7\}$.

Among the closed sets in the last description, some are obviously too big. To rule them out, we need a definition. A set B is called **irreducible** if whenever $i, j \in B$, i communicates with j. The irreducible closed sets in the example are $\{1,5\}$ and $\{4,6,7\}$. The next result explains our interest in irreducible closed sets.

(3.4) Theorem. *If C is a finite closed and irreducible set, then all states in C are recurrent.*

Before entering into an explanation of this result, we note that (3.4) tells us that 1, 5, 4, 6, and 7 are recurrent, completing our study of the Example 3.3 with the results we had claimed earlier.

In fact, the combination of (3.2) and (3.4) is sufficient to classify the states in any finite state Markov chain. An algorithm will be explained in the proof of the following result.

(3.5) Theorem. *If the state space S is finite, then S can be written as a disjoint union $T \cup R_1 \cup \cdots \cup R_k$, where T is a set of transient states and the R_i, $1 \le i \le k$, are closed irreducible sets of recurrent states.*

Proof. Let T be the set of x for which there is a y so that $x \to y$ but $y \not\to x$. The states in T are transient by (3.2). Our next step is to show that all the remaining states, $S - T$, are recurrent.

Pick an $x \in S - T$ and let $C_x = \{y : x \to y\}$. Since $x \in S - T$ it has the property if $x \to y$, then $y \to x$. To check that C_x is closed note that if $y \in C_x$ and $y \to z$, then Exercise 3.1 implies $x \to z$ so $z \in C_x$. To check irreducibility, note that if $y, z \in C_x$, then $x \in S - T$ implies $y \to x$ and we have $x \to z$ by definition, so Exercise 3.1 implies $y \to z$. C_x is closed and irreducible so all states in C_x are recurrent. Let $R_1 = C_x$. If $S - T - R_1 = \emptyset$, we are done. If not, pick a site $w \in S - T - R_1$ and repeat the procedure. □

The rest of this section is devoted to the proof of (3.4).

Why is (3.4) true? The first step to establish a "solidarity" result for two states that communicate, i.e., one state is recurrent if and only if the other is.

46   Chapter 1   Markov Chains

This is the first of several instances of such results that we will see in our study of Markov chains.

(3.6) Lemma. *If x is recurrent and $x \to y$, then y is recurrent.*

Why is this true? Every time the chain returns to x it has a positive probability of hitting y before it returns to x. Thus it will not return to x infinitely many times without hitting y at least once. □

With (3.6) in hand, (3.4) can be proved by showing.

(3.7) Lemma. *In a finite closed set there has to be one recurrent state.*

Why is this true? Since the set is closed, the chain will spend an infinite amount of time in the set. If the chain spent a finite amount of time at each of the finitely many states in the set, we would have a contradiction. □

Proofs of (3.6) and (3.7). To make the arguments above for (3.6) and (3.7) precise we need to introduce a little more theory. We define the time of the kth visit to y by

$$T_y^k = \min\{n > T_y^{k-1} : X_n = y\}$$

for $k \geq 1$ and set $T_y^0 = 0$ to get the sequence of definitions started. In this case, $T_y^1 = \min\{n > 0 : X_n = y\}$ is the return time we called T_y at the beginning of this section.

Let $\rho_{xy} = P_x(T_y < \infty)$ be the probability we ever visit y at some time $n \geq 1$ when we start from x. Here the clumsy phrase "at some time $n \geq 1$" is needed so that the interpretation is correct when $x = y$. It is immediate from the strong Markov property that

(3.8) Lemma. *For $k \geq 1$, $P_x(T_y^k < \infty) = \rho_{xy} \rho_{yy}^{k-1}$.*

Proof. In order to make k visits to y we first have to go from x to y and then return $k - 1$ times from y to y. The strong Markov property implies that the probability all the events occur is the product of their probabilities. □

Let $N(y)$ be the number of visits to y at times $n \geq 1$. Using (3.8) we can compute $EN(y)$.

(3.9) Lemma.
$$E_x N(y) = \frac{\rho_{xy}}{1 - \rho_{yy}}$$

Section 1.3 Classification of States

Proof. Recall that for any nonnegative integer valued random variable N, the expected value of N can be computed by

$$(3.10) \quad EN = \sum_{k=1}^{\infty} P(N \geq k)$$

Now the probability of returning at least k times, $\{N \geq k\}$, is the same as the event that the kth return occurs, i.e., $\{T_y^k < \infty\}$, so using (3.8) we have

$$E_x N(y) = \rho_{xy} \sum_{k=1}^{\infty} \rho_{yy}^{k-1} = \frac{\rho_{xy}}{1 - \rho_{yy}}$$

since $\sum_{n=0}^{\infty} \theta^n = 1/(1-\theta)$ whenever $|\theta| < 1$. □

Proof of (3.10). Let $i_N(k) = 1$ if $N \geq k$ and 0 otherwise. It is easy to see that $N = \sum_{k=1}^{\infty} i_N(k)$. Taking expected values and noticing $Ei_N(k) = P(N \geq k)$ gives

$$EN = \sum_{k=1}^{\infty} Ei_N(k) = \sum_{k=1}^{\infty} P(N \geq k) \qquad \square$$

Our next step is to compute the expected number of returns to y in a different way.

(3.11) Lemma. $E_x N(y) = \sum_{n=1}^{\infty} p^n(x,y)$.

Proof. Let $1_{\{X_n=y\}}$ denote the random variable that is 1 if $X_n = y$, 0 otherwise. Clearly $N(y) = \sum_{n=1}^{\infty} 1_{\{X_n=y\}}$. Taking expected values now gives

$$E_x N(y) = \sum_{n=1}^{\infty} P_x(X_n = y) \qquad \square$$

With the two lemmas established we can now state our next main result.

(3.12) Theorem. y is recurrent if and only if

$$\sum_{n=1}^{\infty} p^n(y,y) = E_y N(y) = \infty$$

Proof. The first equality is (3.11). From (3.9) we see that $E_y N(y) = \infty$ if and only if $\rho_{yy} = 1$, which is the definition of recurrence. □

The excitement about this result is that it rules out a middle ground between recurrence and transience. If y is recurrent and we start at y, then $N(y) = \infty$ with probability one, while if y is transient, then $N(y) < \infty$ with probability one and in fact $N(y)$ has a finite expected value. We will see later that (3.12) is useful in treating concrete examples. For the moment we need it to complete the proofs of (3.6) and (3.7).

Proof of (3.6). Suppose x is recurrent and $\rho_{xy} > 0$. By the first sentence of the proof of (3.2) we must have $\rho_{yx} > 0$. Pick j and ℓ so that $p^j(y, x) > 0$ and $p^\ell(x, y) > 0$. $p^{j+k+\ell}(y, y)$ is probability of going from y to y in $j + k + \ell$ steps while the product $p^j(y, x) p^k(x, x) p^\ell(x, y)$ is the probability of doing this and being at x at times j and $j + k$. Thus we must have

$$\sum_{k=0}^{\infty} p^{j+k+\ell}(y, y) \geq p^j(y, x) \left(\sum_{k=0}^{\infty} p^k(x, x) \right) p^\ell(x, y)$$

If x is recurrent then $\sum_k p^k(x, x) = \infty$, so $\sum_m p^m(y, y) = \infty$ and (3.12) implies that y is recurrent. □

Proof of (3.7). If all the states in C are transient then (3.9) implies that $E_x N(y) < \infty$ for all x and y in C. Since C is finite it follows that

$$\infty > \sum_{y \in C} E_x N(y) = E_x \sum_{y \in C} N(y)$$

However, this is impossible. C is closed and the chain stays in it for an infinite amount of time, so $\sum_{y \in C} N(y) = \infty$ with probability one. □

1.4. Limit Behavior

If y is a transient state, then X_n will only return to y finitely many times. From this it follows that the probability of a return after time n goes to 0 as $n \to \infty$, and hence

$$p^n(x, y) = P_x(X_n = y) \to 0 \quad \text{for any initial state } x$$

We will see that in most cases if y is a recurrent state in a finite state chain, then $p^n(x, y)$ will converge to a positive limit. Our first example shows one problem that can prevent convergence.

Section 1.4 Limit Behavior   49

Example 4.1. Ehrenfest chain. For concreteness, suppose there are three balls. In this case the transition probability is

	0	1	2	3
0	0	3/3	0	0
1	1/3	0	2/3	0
2	0	2/3	0	1/3
3	0	0	3/3	0

In the second power of p the zero pattern is shifted:

	0	1	2	3
0	1/3	0	2/3	0
1	0	7/9	0	2/9
2	2/9	0	7/9	0
3	0	2/3	0	1/3

To see that the zeros will persist, note that if initially we have an odd number of balls in the left urn, then no matter whether we add or subtract one the result will be an even number. Likewise, if the number is even initially, then it will be odd after one step. This alternation between even and odd means that it is impossible to be back where we started after an odd number of steps. In symbols, if n is odd then $p^n(x, x) = 0$ for all x. □

To see that the problem in the last example can occur for multiples of any number N consider:

Example 4.2. Repair chain. A machine has three critical parts that are subject to failure, but can function as long as two of these parts are working. When two are broken, they are replaced and the machine is back to working order the next day. To formulate a Markov chain model we declare its state space to be the parts that are broken $\{0, 1, 2, 3, 12, 13, 23\}$. This time we assume that each day one part chosen at random breaks to arrive at the following transition matrix:

	0	1	2	3	12	13	23
0	0	1/3	1/3	1/3	0	0	0
1	0	0	0	0	1/2	1/2	0
2	0	0	0	0	1/2	0	1/2
3	0	0	0	0	0	1/2	1/2
12	1	0	0	0	0	0	0
13	1	0	0	0	0	0	0
23	1	0	0	0	0	0	0

Clearly, in this case $p^n(x,x) = 0$ unless n is a multiple of 3. By considering a machine with N parts we can get an example where $p^n(x,x) = 0$ unless n is a multiple of N.

Definition. The **period** of a state is the largest number that will divide all the n for which $p^n(x,x) > 0$. That is, it is the greatest common divisor of $I_x = \{n \geq 1 : p^n(x,x) > 0\}$.

To check that this definition works correctly, we note that in Example 4.1, $\{n \geq 1 : p^n(x,x) > 0\} = \{2, 4, \ldots\}$, so the greatest common divisor is 2. Similarly, in Example 4.2, $\{n \geq 1 : p^n(x,x) > 0\} = \{N, 2N, \ldots\}$, so the greatest common divisor is N.

EXERCISE 4.1. Show that I_x is closed under addition. That is, if $i, j \in I_x$, then $i + j \in I_x$. Repeatedly applying this gives that if $i_1, \ldots i_m \in I_x$ and $a_1, \ldots a_m$ are nonnegative integers, then $a_1 i_1 + \cdots + a_m i_m \in I_x$.

While periodicity is a theoretical possibility, it rarely manifests itself in applications, except occasionally as an odd-even parity problem, e.g., the Ehrenfest chain. In most cases we will find (or design) our chain to be **aperiodic**, i.e., all states have period 1. To be able to verify this property for examples, we need to discuss some theory.

(4.1) Lemma. If $p(x,x) > 0$, then x has period 1.

Proof. If $p(x,x) > 0$, then $1 \in I_x$, so the greatest common divisor must be 1.□

This fact helps us easily settle the question for

Example 4.3. Weather chain.

	1	2	3
1	.4	.6	0
2	.2	.5	.3
3	.1	.7	.2

we have $p(x,x) > 0$ for all x, so all states have period 1.

While $p(x,x) > 0$ is sufficient for x to have period 1, it is not necessary.

Example 4.4. Triangle and square. Consider the transition matrix:

	-2	-1	0	1	2	3
-2	0	0	1	0	0	0
-1	1	0	0	0	0	0
0	0	0.5	0	0.5	0	0
1	0	0	0	0	1	0
2	0	0	0	0	0	1

Pictorially, this may be described as

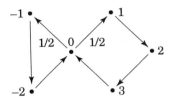

In words, from 0 we are equally likely to go to -1 or to 1. If we go to -1, we continue with $-1 \to -2 \to 0$, returning in 3 steps. If we go to 1, then we continue with $1 \to 2 \to 3 \to 0$, returning in 4 steps. Since $3, 4 \in I_x$ and the greatest common divisor of 3 and 4 is 1, it follows that the greatest common divisor of I_x is 1; i.e., x has period 1.

To motivate the next result we note that in Example 4.4 by going around twice we can return to 0 in 6, 7, or 8 steps. With three trips the possibilities are 9, 10, 11, or 12, and with four trips the possibilities are any number from 12 to 16, etc. Thus

$$I_0 = \{3, 4, 6, 7, 8, 9, 10, 11, 12, \ldots\}$$

(4.2) Lemma. *If x has period 1, then there is a number n_0 so that if $n \geq n_0$, then $n \in I_x$.*

In words, I_x contains all of the integers after some value n_0. In Example 4.4 above, $n_0 = 6$.

Proof. We begin by observing that it enough to show that I_x will contain two consecutive integers: k and $k + 1$. For then it will contain $2k, 2k + 1, 2k + 2$, or in general $jk, jk + 1, \ldots jk + j$. For $j \geq k - 1$ these blocks overlap and no integers are left out.

To show that there are two consecutive integers, we cheat and use a fact from number theory: if the greatest common divisor of a set I_x is 1 then there are integers $i_1, \ldots i_m \in I_x$ and (positive or negative) integer coefficients c_i so that $c_1 i_1 + \cdots + c_m i_m = 1$. Let $a_i = c_i^+$ and $b_i = (-c_i)^+$. In words the a_i are the positive coefficients and the b_i are -1 times the negative coefficients. Rearranging the last equation gives

$$a_1 i_1 + \cdots + a_m i_m = (b_1 i_1 + \cdots + b_m i_m) + 1$$

and using Exercise 4.1 we have found our two consecutive integers in I_x. □

To explain why the proof is mysterious consider the following:

Example 4.5. Mathematician's nightmare. Here, $\{0, 1, \ldots, 14\}$ is the state space. From 0 the chain goes to 5, 9, or 14 with probability $1/3$. If $x > 0$, then $p(x, x-1) = 1$, so depending on where the chain lands it will return to 0 in 6, 10, or 15 steps. Here I_0 contains $6 = 2 \cdot 3$, $10 = 2 \cdot 5$, and $15 = 3 \cdot 5$, so the greatest common divisor is 1. Fifteen and $16 = 6 + 10$ are consecutive integers in I_0, so the proof guarantees that all numbers beyond $14 \cdot 15 = 210$ will be covered. A little calculation shows

$$I_0 = \{6, 10, 15, 16, 20, 21, 22, 25, 26, 27, 28, 30, 31, 32, 33, 34, 35, \ldots\}$$

so in this case $n_0 = 30$.

For many examples it is useful to know the following:

(4.3) Lemma. *If x and y communicate with each other ($x \to y$ and $y \to x$), then x and y have the same period.*

Why is this true? The short answer is that if the two states try to have different periods, then by going from x to y, from y to y in the various possible ways, and then from y to x, we will get a contradiction.

Proof. Suppose that the period of x is c, while the period of y is $d < c$. Let k be such that $p^k(x, y) > 0$ and let m be such that $p^m(y, x) > 0$. Since

$$p^{k+m}(x, x) \geq p^k(x, y) p^m(y, x) > 0$$

we have $k + m \in I_x$. Since x has period c, $k + m$ must be a multiple of c. Now let ℓ be any integer with $p^\ell(y, y) > 0$. Since

$$p^{k+\ell+m}(x, x) \geq p^k(x, y) p^\ell(y, y) p^m(y, x) > 0$$

$k+\ell+m \in I_x$, and $k+\ell+m$ must be a multiple of c. Since $k+m$ is itself a multiple of c, this means that ℓ is a multiple of c. Since $\ell \in I_y$ was arbitrary, we have shown that c is a divisor of every element of I_y, but $d < c$ is the greatest common divisor, so we have a contradiction. □

To illustrate the usefulness of (4.3) we consider:

Example 4.6. Symmetric reflecting random walk on the line. The state space is $\{0, 1, \ldots, N\}$. The chain goes to the right or left at each step with probability $1/2$, subject to the rules that if it tries to go to the left from 0 or to the right from N it stays put. For example, when $N = 4$ the transition probability is

	0	1	2	3	4
0	0.5	0.5	0	0	0
1	0.5	0	0.5	0	0
2	0	0.5	0	0.5	0
3	0	0	0.5	0	0.5
4	0	0	0	0.5	0.5

Looking at the matrix we see that $p(0, 0) > 0$, so 0 has period 1. Since all states communicate with 0 it follows that all states have period 1.

Our first step in studying the limiting behavior of Markov chains was to describe the condition ("periodicity") that could prevent convergence. Our second step is to identify the limit

$$\lim_{n \to \infty} p^n(x, y) = \pi(y)$$

A **stationary distribution** π is a solution of $\pi p = \pi$ with $\sum_y \pi(y) = 1$. Here we are regarding π as a row vector ($1 \times k$ matrix) and multiplying the matrix p on the left. In summation notion this can be written as

$$\sum_x \pi(x) p(x, y) = \pi(y)$$

One of the reasons for our interest in stationary distributions is explained by:

(4.4) Lemma. If X_0 has distribution π, i.e., $P(X_0 = y) = \pi(y)$, then so does X_n for all $n \geq 1$.

In words, the stationary distribution is an equilibrium state for the chain. If the chain starts with this distribution at time 0, then it will have this distribution at all times.

54   Chapter 1   Markov Chains

Proof of (4.4). Formula (2.3) implies

$$P(X_n = y) = (\pi p^n)(y)$$

i.e., we multiply the row vector π by the p^n, the nth power of the matrix p and look at the y component. Since $\pi p = \pi$ it follows that

$$\pi p^n = (\pi p)p^{n-1} = \pi p^{n-1} = (\pi p)p^{n-2} = \pi p^{n-2} \ldots = \pi \qquad \square$$

We now come to the main results of the section. The proofs of (4.5), (4.7), and (4.8) are somewhat complicated, so we have delayed them to Section 1.8.

(4.5) Convergence theorem. *Suppose p is irreducible, aperiodic, and has a stationary distribution π. Then as $n \to \infty$, $p^n(x, y) \to \pi(y)$.*

To begin to digest this result we start by seeing what it says for:

Example 4.7. Weather chain.

	1	2	3
1	.4	.6	0
2	.2	.5	.3
3	.1	.7	.2

The equation $\pi p = \pi$ says

$$\pi_1(.4) + \pi_2(.2) + \pi_3(.1) = \pi_1$$
$$\pi_1(.6) + \pi_2(.5) + \pi_3(.7) = \pi_2$$
$$\pi_2(.3) + \pi_3(.2) = \pi_3$$

Here we have three equations in three unknowns. In addition there is a fourth relationship: $\pi_1 + \pi_2 + \pi_3 = 1$, so one of the equations is redundant.

If we set $\pi_3 = c$, then the third equation implies $\pi_2 = 8c/3$. Plugging this into the second equation, we have

$$\pi_1 = \frac{10}{6}\left(\frac{5}{10}\pi_2 - \frac{7}{10}\pi_3\right) = \frac{5}{6} \cdot \frac{8c}{3} - \frac{7}{6} \cdot c = \frac{19c}{18}$$

Since $\pi_1 + \pi_2 + \pi_3 = 1$, we must have $1 = (19 + 48 + 18)c/18 = 85c/18$ or $c = 18/85$. This means that the stationary distribution is given by

$$\pi_1 = \frac{19}{85} = .22353, \qquad \pi_2 = \frac{48}{85} = .56471, \qquad \pi_3 = \frac{18}{85} = .21176$$

Using (4.5) now we see that $\lim_{n\to\infty} p^n(i,j) = \pi_j$. As we saw in Example 2.1, n does not have to be very large for $p^n(i,j)$ to be close to its limiting value.

$$p^4 = \begin{pmatrix} .2278 & .5634 & .2088 \\ .2226 & .5653 & .2121 \\ .2215 & .5645 & .2140 \end{pmatrix}$$

$$p^8 = \begin{pmatrix} .22355 & .56470 & .21175 \\ .22352 & .56471 & .21177 \\ .22352 & .56471 & .21177 \end{pmatrix}$$

Looking at the last column of p^8 we see that $p^8(i,3) \approx .21176$ for $i = 1, 2, 3$. Another way of saying this is that (in our Markov chain model) the weather today has almost no effect on the weather eight days from now. □

Existence and uniqueness of stationary distributions. The statement of (4.5) naturally suggests two questions: "When is there a stationary distribution?" and "Can there be more than one?" Starting with the easier second question we see that from (4.5) we immediately get:

(4.6) Corollary. *If p is irreducible, then the stationary distribution is unique. That is, it is impossible to have more than 1, but we might have 0.*

Proof. First suppose that p is aperiodic. If there were two stationary distributions, π_1 and π_2, then by applying (4.5) we would conclude that

$$\pi_1(y) = \lim_{n\to\infty} p^n(x,y) = \pi_2(y)$$

To get rid of the aperiodicity assumption, let I be the transition probability for the chain that never moves, i.e., $I(x,x) = 1$ for all x, and define a new transition probability $\hat{p} = (I + p)/2$, i.e., we either do nothing with probability $1/2$ or take a step according to p. Since $p(x,x) \geq 1/2$ for all x, $\hat{p}$ is aperiodic. Now, if $\pi p = \pi$ then $\pi\hat{p} = \pi$, so $\hat{p}$ has at least as many stationary distributions as p does. However, $\hat{p}$ has at most 1, so p can have at most 1. □

Only later in (7.3) will we be able to give necessary and sufficient conditions for the existence of a stationary distribution. Until then the reader can rest assured that the following is true:

(4.7) Theorem. *If the state space S is finite then there is at least one stationary distribution.*

A second important fact about the stationary distribution is that it gives us the "limiting fraction for time we spend in each state." This property is true even for periodic chains. For convenience we derive this from a more general fact.

(4.8) **Strong law for Markov chains.** *Suppose p is irreducible and has stationary distribution π. Let $r(x)$ be the reward we earn in state x and suppose that $\sum_x \pi(x)|r(x)| < \infty$. Then as $n \to \infty$ we have*

$$\frac{1}{n}\sum_{k=1}^{n} r(X_k) \to \sum_x \pi(x)r(x)$$

To explain the name recall that if $X_1, X_2, \ldots$ are independent with distribution $\pi(x)$, then the strong law of large numbers applied to $r(X_1), r(X_2), \ldots$ gives

$$\frac{1}{n}\sum_{k=1}^{n} r(X_k) \to Er(X_1) = \sum_x \pi(x)r(x)$$

If we let $r(y) = 1$ and $r(x) = 0$ for $x \neq y$, then $\sum_{k=1}^{n} r(X_k)$ gives the number of visits to y up to time n, $N_n(y)$, and the result in (4.8) becomes

(4.9) **Corollary.** *As $n \to \infty$, $N_n(y)/n \to \pi(y)$.*

Specializing further to the weather chain and y the sunny state, this says that if you keep track of the number of sunny days, and divide by the number of days, n, for which you have data, then in the limit as $n \to \infty$ your observed fraction of sunny days will converge to 18/85.

To illustrate the use of the stronger statement in (4.8) for reward functions, we return to our study of Example 1.4.

Example 4.8. Repair chain. A machine has three critical parts that are subject to failure, but can function as long as two of these parts are working. When two are broken, they are replaced and the machine is back to working order the next day. Declaring the state space to be the parts that are broken $\{0, 1, 2, 3, 12, 13, 23\}$, we arrived in Example 1.4 at a Markov chain X_n with the following transition matrix:

	0	1	2	3	12	13	23
0	.93	.01	.02	.04	0	0	0
1	0	.94	0	0	.02	.04	0
2	0	0	.95	0	.01	0	.04
3	0	0	0	.97	0	.01	.02
12	1	0	0	0	0	0	0
13	1	0	0	0	0	0	0
23	1	0	0	0	0	0	0

Section 1.4 Limit Behavior

and we asked: If we are going to operate the machine for 1800 days (about 5 years) then how many parts of types 1, 2, and 3 will we use?

Solution. Skipping the first column, the equations $\pi p = \pi$ imply

$$.01\pi(0) + .94\pi(1) = \pi(1) \qquad .02\pi(1) + .01\pi(2) = \pi(12)$$
$$.02\pi(0) + .95\pi(2) = \pi(2) \qquad .04\pi(1) + .01\pi(3) = \pi(13)$$
$$.04\pi(0) + .97\pi(3) = \pi(3) \qquad .04\pi(2) + .02\pi(3) = \pi(23)$$

Starting with $\pi(0) = c$ we find $\pi(1) = c/6 = 5c/30$, $\pi(2) = 2c/5 = 12c/30$, $\pi(3) = 4c/3 = 40c/30$, then

$$\pi(12) = (10c + 12c)/3000$$
$$\pi(13) = (20c + 40c)/3000$$
$$\pi(23) = (48c + 80c)/3000$$

Summing the π's gives $(3000 + 500 + 1200 + 4000 + 22 + 60 + 128)c/3000 = 8910c/3000$ Thus $c = 3000/8910$ and we have

$$\pi(0) = 3000/8910$$
$$\pi(1) = 500/8910 \quad \pi(2) = 1200/8910 \quad \pi(3) = 4000/8910$$
$$\pi(12) = 22/8910 \quad \pi(13) = 60/8910 \quad \pi(23) = 128/8910$$

We use up one part of type 1 on each visit to 12 or to 13, so on the average we use 82/8910 of a part per day. Over 1800 days we will use an average of $1800 \cdot 82/8910 = 16.56$ parts of type 1. Similarly type 2 and type 3 parts are used at the long run rates of 150/8910 and 188/8910 per day, so over 1800 days we will use an average of 30.30 parts of type 2 and 37.98 parts of type 3. □

With a little more work (4.8) can be used to study Example 1.5.

Example 4.9. Inventory chain. We have an electronics store that sells a videogame system, selling 0, 1, 2, or 3 of these units each day with probabilities .3, .4, .2, and .1. Each night at the close of business new units can be ordered which will be available when the store opens in the morning. Suppose that sales produce a profit of $12 but it costs $2 a day to keep unsold units in the store overnight. Since it is impossible to sell 4 units in a day, and it costs us to have unsold inventory we should never have more than 3 units on hand.

Suppose we use a 2,3 inventory policy. That is, we order if there are ≤ 2 units and we order enough stock so that we have 3 units at the beginning of the next day. In this case we always start the day with 3 units, so the transition probability has constant rows

	0	1	2	3
0	.1	.2	.4	.3
1	.1	.2	.4	.3
2	.1	.2	.4	.3
3	.1	.2	.4	.3

In this case it is clear that the stationary distribution is $\pi(0) = .1$, $\pi(1) = .2$, $\pi(2) = .4$, and $\pi(3) = .3$. If we end the day with k units then we sold $3 - k$ and have to keep k over night. Thus our long run sales under this scheme are

$$.1(36) + .2(24) + .4(12) = 3.6 + 4.8 + 4.8 = 13.2 \quad \text{dollars per day}$$

while the inventory holding costs are

$$2(.2) + 4(.4) + 6(.3) = .4 + 1.6 + 1.8 = 3.8$$

for a net profit of 9.4 dollars per day.

Suppose we use a 1,3 inventory policy. In this case the transition probability is

	0	1	2	3
0	.1	.2	.4	.3
1	.1	.2	.4	.3
2	.3	.4	.3	0
3	.1	.2	.4	.3

From the first of the equations for the stationary distribution we get

$$\pi(0) = .1\pi(0) + .1\pi(1) + .3\pi(2) + .1\pi(3) = .1 + .2\pi(2)$$

since the sum of the π's is 1. Similar reasoning shows that

$$\pi(1) = .2 + .2\pi(2), \quad \pi(2) = .4 - .1\pi(2), \quad \pi(3) = .3 - .3\pi(2)$$

The middle equation implies that $\pi(2) = 4/11$, which then gives us

$$\pi(0) = 1/10 + .8/11 = 19/110$$
$$\pi(1) = 2/10 + .8/11 = 30/110$$
$$\pi(2) = 40/110$$
$$\pi(3) = 3/10 - 1.2/11 = 21/110$$

To compute the profit we make from sales note that if we always had enough stock then we would make

$$.1(36) + .2(24) + .4(12) = 13.2 \quad \text{dollars per day}$$

as in the first case. However, when $X_n = 2$ and the demand is 3, an event with probability $(4/11) \cdot .1 = .036$, we lose exactly one of our sales. From this it follows that in the long run we make a profit of

$$13.2 - (.036)12 = 12.768 \quad \text{dollars per day}$$

Our inventory holding cost under the new system is

$$2 \cdot \frac{30}{110} + 4 \cdot \frac{40}{110} + 6 \cdot \frac{21}{110} = \frac{60 + 160 + 126}{110} = 3.145$$

so now our profits are $12.768 - 3.145 = 9.623$.

EXERCISE 4.2. Compute the long-run profit from the 0,3 inventory policy.

1.5. Some Special Examples

A transition matrix p is said to be **doubly stochastic** if its COLUMNS sum to 1, or in symbols $\sum_x p(x,y) = 1$. The adjective "doubly" refers to the fact that by its definition a transition probability matrix has ROWS that sum to 1, i.e., $\sum_y p(x,y) = 1$. The stationary distribution is easy to guess in this case:

(5.1) Theorem. *If p is a doubly stochastic transition probability for a Markov chain with N states, then the uniform distribution, $\pi(x) = 1/N$ for all x, is a stationary distribution.*

Proof. To check this claim we note that

$$\sum_x \pi(x) p(x,y) = \frac{1}{N} \sum_x p(x,y) = \frac{1}{N}$$

so the uniform distribution $\pi(x) = 1/N$ for all x satisfies $\pi p = \pi$. Looking at the second equality we see that conversely, if the stationary distribution is uniform then p is doubly stochastic. □

One doubly stochastic chain we have already seen in Example 4.6 is

Example 5.1. Symmetric reflecting random walk on the line. The state space is $\{0, 1, 2 \ldots, L\}$. The chain goes to the right or left at each step with probability 1/2, subject to the rules that if it tries to go to the left from 0 or to the right from L it stays put. For example, when $L = 4$ the transition probability is

	0	1	2	3	4
0	0.5	0.5	0	0	0
1	0.5	0	0.5	0	0
2	0	0.5	0	0.5	0
3	0	0	0.5	0	0.5
4	0	0	0	0.5	0.5

It is clear in the example $L = 4$ that each column adds up to 1. With a little thought one sees that this is true for any L, so the stationary distribution is uniform, $\pi(i) = 1/(L+1)$. □

For a new example of a doubly stochastic transition probability consider:

Example 5.2. Nearest-neighbor random walk on a circle. The state space is $\{0, 1, \ldots, L\}$. At each step, the chain goes to the right with probability a or to the left with probability $1 - a$, subject to the rules that if it tries to go to the left from 0 it ends up at L or if tries to go to the right from L it ends up at 0. For example, when $L = 4$ the transition probability is

	0	1	2	3	4
0	0	a	0	0	$1-a$
1	$1-a$	0	a	0	0
2	0	$1-a$	0	a	0
3	0	0	$1-a$	0	a
4	a	0	0	$1-a$	0

Again, it is clear in the example $L = 4$ that each column adds up to 1, and with a little thought one sees that this is true for any L, so the stationary distribution is uniform, $\pi(i) = 1/(L+1)$.

EXERCISE 5.1. Show that the chain is always irreducible but aperiodic if and only if L is even.

EXERCISE 5.2. Consider the case $a = 2/3$ and $L = 5$. Computing p^2 and rearranging the entries, we have

	0	2	4	1	3	5
0	5/9	4/9	1/9	0	0	0
2	1/9	5/9	4/9	0	0	0
4	4/9	1/9	5/9	0	0	0
1	0	0	0	5/9	4/9	1/9
3	0	0	0	1/9	5/9	4/9
5	0	0	0	4/9	1/9	5/9

(a) Use (4.5) to find $\lim_{n\to\infty} p^{2n}(x,y)$. (b) Use the result from (a) and $p^{2n+1} = p^{2n}p$ to find $\lim_{n\to\infty} p^{2n+1}(x,y)$.

Birth and death chains will be our second set of examples. These chains are defined by the property that the state space is some sequence of integers $\ell, \ell+1, \ldots r-1, r$ and it is impossible to jump by more than one:

$$p(x,y) = 0 \quad \text{when } |x-y| > 1$$

One of the special features that makes this class of examples easy to study is that the stationary distribution satisfies the

(5.2) Detailed balance condition. $\pi(x)p(x,y) = \pi(y)p(y,x)$

To see that this is a stronger condition than $\pi p = \pi$, we sum over x on each side to get

$$\sum_x \pi(x)p(x,y) = \pi(y)\sum_x p(y,x) = \pi(y)$$

Pictorially, we think of $\pi(x)$ as giving the amount of sand at x, and one transition of the chain as sending a fraction $p(x,y)$ of the sand at x to y. In this case the detailed balance condition says that the amount of sand going from x to y in one step is exactly balanced by the amount going back from y to x. In contrast the condition $\pi p = \pi$ says that after all the transfers are made, the amount of sand that ends up at each site is the same as the amount that starts there.

Many chains do not have stationary distributions that satisfy the detailed balance condition.

Example 5.3. Weather chain. In this case, we have $p(1,3) = 0$ but all other $p(x,y) > 0$. Any chain with this property cannot have a stationary distribution that satisfies (5.2) since

$$0 = \pi(1)p(1,3) = \pi(3)p(3,1) \quad \text{implies } \pi(3) = 0$$
$$\text{then } 0 = \pi(3)p(3,2) = \pi(2)p(2,3) \quad \text{implies } \pi(2) = 0$$
$$\text{and } 0 = \pi(2)p(2,1) = \pi(1)p(1,2) \quad \text{implies } \pi(1) = 0$$

and we arrive at the conclusion that $\pi(x) = 0$ for all x. Generalizing from this example we can see that there can be no stationary distribution with detailed balance unless $p(x,y) > 0$ implies that $p(y,x) > 0$.

Some good news is that birth and death chains always have stationary distributions that satisfy detailed balance condition and this makes their stationary distributions very easy to compute. We study a special case first then derive a general formula.

Example 5.4. Ehrenfest chain. For concreteness, suppose there are three balls. In this case the transition probability is

	0	1	2	3
0	0	3/3	0	0
1	1/3	0	2/3	0
2	0	2/3	0	1/3
3	0	0	3/3	0

Drawing the graph of the chain.

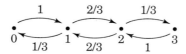

We see that the detailed balance equations say:

$$\pi(0) = \pi(1)/3 \qquad 2\pi(1)/3 = 2\pi(2)/3 \qquad \pi(2)/3 = \pi(3)$$

Setting $\pi(0) = c$ we can solve to get $\pi(1) = 3c$, $\pi(2) = \pi(1) = 3c$, and $\pi(3) = c$. The sum of the π's should be one, so we pick $c = 1/8$ to get

$$\pi(0) = 1/8, \qquad \pi(1) = 3/8, \qquad \pi(2) = 3/8, \qquad \pi(3) = 1/8$$

Knowing the answer in general, one can look at the last equation and see that π represents the distribution of the number of Heads when we flip three coins, then guess in general that the binomial distribution with $p = 1/2$ is the stationary distribution:

(5.3) $$\pi(i) = 2^{-n} \binom{n}{i}$$

Here $m! = 1 \cdot 2 \cdots (m-1) \cdot m$, with $0! = 1$, and

$$\binom{n}{x} = \frac{n!}{x!(n-x)!}$$

is the binomial coefficient which gives the number of ways of choosing x objects out of a set of n. To check that our guess satisfies the detailed balance condition, we note that

$$\pi(x)p(x,x+1) = 2^{-n}\frac{n!}{x!(n-x)!} \cdot \frac{n-x}{n}$$

$$= 2^{-n}\frac{n!}{(x+1)!(n-x-1)!} \cdot \frac{x+1}{n} = \pi(x+1)p(x+1,x)$$

General birth and death chains. Suppose that the state space is $\{\ell, \ell+1, \ldots, r-1, r\}$ and the transition probability has

$$\begin{aligned} p(x,x+1) &= p_x & \text{for } x < r \\ p(x,x-1) &= q_x & \text{for } x > \ell \\ p(x,x) &= r_x & \text{for } \ell \leq x \leq r \end{aligned}$$

while the other $p(x,y) = 0$. If $x < r$ detailed balance between x and $x+1$ implies $\pi(x)p_x = \pi(x+1)q_{x+1}$, so

(5.4) $$\pi(x+1) = \frac{p_x}{q_{x+1}} \cdot \pi(x)$$

Using this with $x = \ell$ gives $\pi(\ell+1) = \pi(\ell)p_\ell/q_{\ell+1}$. Taking $x = \ell+1$

$$\pi(\ell+2) = \frac{p_{\ell+1}}{q_{\ell+2}} \cdot \pi(\ell+1) = \frac{p_{\ell+1} \cdot p_\ell}{q_{\ell+2} \cdot q_{\ell+1}} \cdot \pi(\ell)$$

Extrapolating from the first two results we see that in general

(5.5) $$\pi(\ell+i) = \pi(\ell) \cdot \frac{p_{\ell+i-1} \cdot p_{\ell+i-2} \cdots p_{\ell+1}p_\ell}{q_{\ell+i} \cdot q_{\ell+i-1} \cdots q_{\ell+2} \cdots q_{\ell+1}}$$

To keep the indexing straight note that: (i) there are i terms in the numerator and in the denominator, (ii) the indices decrease by 1 each time, (iii) the answer will not depend on $p_{\ell+i}$ or q_ℓ.

For a concrete example to illustrate the use of this formula consider

Example 5.5. Three machines, one repairman. Suppose that an office has three machines that each break with probability .1 each day, but when there is at least one broken, then with probability 0.5 the repairman can fix one of them for use the next day. If we ignore the possibility of two machines breaking on the same day, then the number of working machines can be modeled as a birth and death chain with the following transition matrix:

	0	1	2	3
0	.5	.5	0	0
1	.05	.5	.45	0
2	0	.1	.5	.4
3	0	0	.3	.7

Rows 0 and 3 are easy to see. To explain row 1, we note that the state will only decrease by 1 if one machine breaks and the repairman fails to repair the one he is working on, an event of probability $(.1)(.5)$, while the state can only increase by 1 if he succeeds and there is no new failure, an event of probability $.5(.9)$. Similar reasoning shows $p(2,1) = (.2)(.5)$ and $p(2,3) = .5(.8)$.

To find the stationary distribution we use the recursive formula (5.4) to conclude that if $\pi(0) = c$ then

$$\pi(1) = \pi(0) \cdot \frac{p_0}{q_1} = c \cdot \frac{.5}{.05} = 10c$$

$$\pi(2) = \pi(1) \cdot \frac{p_1}{q_2} = 10c \cdot \frac{.45}{.1} = 45c$$

$$\pi(3) = \pi(2) \cdot \frac{p_2}{q_3} = 45c \cdot \frac{.4}{.3} = 60c$$

The sum of the π's is $116c$, so if we let $c = 1/116$ then we get

$$\pi(3) = \frac{60}{116}, \quad \pi(2) = \frac{45}{116}, \quad \pi(1) = \frac{10}{116}, \quad \pi(0) = \frac{1}{116} \qquad \square$$

There are many other Markov chains that are not birth and death chains but have stationary distributions that satisfy the detailed balance condition. A large number of possibilities are provided by

Example 5.6. Random walks on graphs. A graph is described by giving two things: (i) a set of vertices V (which, for the moment, we will suppose is a finite set) and (ii) an adjacency matrix $A(u,v)$, which is 1 if u and v are "neighbors" and 0 otherwise. By convention we set $A(v,v) = 0$ for all $v \in V$. The degree of a vertex u is equal to the number of neighbors it has. In symbols,

$$d(u) = \sum_v A(u,v)$$

since each neighbor of u contributes 1 to the sum. We write the degree this way to make it clear that

(∗) $$p(u,v) = \frac{A(u,v)}{d(u)}$$

defines a transition probability. In words, if $X_n = u$, we jump to a randomly chosen neighbor of u at time $n+1$.

It is immediate from (∗) that if c is a positive constant then $\pi(u) = cd(u)$ satisfies the detailed balance condition:

$$\pi(u)p(u,v) = cA(u,v) = cA(v,u) = \pi(v)p(u,v)$$

Thus, if we take $c = 1/\sum_u d(u)$, we have a stationary probability distribution. For a concrete example, consider

Example 5.7. Random walk on a 4 by 4 checkerboard. Thinking of the two indices as the rows and the columns, we choose the set of vertices V to be

(1,1)	(1,2)	(1,3)	(1,4)
(2,1)	(2,2)	(2,3)	(2,4)
(3,1)	(3,2)	(3,3)	(3,4)
(4,1)	(4,2)	(4,3)	(4,4)

We set $A(u,v) = 1$ if v can be obtained from u by adding or subtracting 1 from either coordinate. For example, $u = (2,1)$ is adjacent to $v = (1,1)$, $(3,1)$, and $(2,2)$. Geometrically two vertices are neighbors if they are adjacent in some row or column. The degrees of the vertices are

2	3	3	2
3	4	4	3
3	4	4	3
2	3	3	2

The sum of the degrees is $4 \cdot 4 + 8 \cdot 3 + 4 \cdot 2 = 16 + 24 + 8 = 48$, so the stationary probabilities are

1/24	1/16	1/16	1/24
1/16	1/12	1/12	1/16
1/16	1/12	1/12	1/16
1/24	1/16	1/16	1/24

EXERCISE 5.3. Suppose we modify the rules so that the particle picks one of four directions to move in and stays put if it is not possible to move in that direction? What is the new stationary distribution?

1.6. One-Step Calculations

The title of this section refers to a technique for computing probabilities and expected values for Markov chains by considering how the situation changes when the chain takes one step.

Example 6.1. Gambler's ruin. Consider a gambling game in which on any turn you win \$1 with probability p or lose \$1 with probability $1-p$. Suppose further that you will quit playing if your fortune reaches \$$N$. Of course, if your fortune reaches \$0, then the casino makes you stop. For reasons that will become clear in a moment, we depart from our usual definition and let

$$V_y = \min\{n \geq 0 : X_n = y\}$$

be the time of the first visit to y. Let

$$h(x) = P_x(V_N < V_0)$$

be the happy event that our gambler reaches the goal of \$$N$ before going bankrupt when starting with \$$x$. Thanks to our definition of V_x as the minimum of $n \geq 0$ we have $h(0) = 0$, and $h(N) = 1$. To calculate $h(i)$ for $0 < i < N$, consider what happens on the first step to arrive at

$$h(i) = ph(i+1) + (1-p)h(i-1)$$

To solve this we rearrange to get $p(h(i+1) - h(i)) = (1-p)(h(i) - h(i-1))$ and conclude

$$(\star) \qquad h(i+1) - h(i) = \frac{1-p}{p} \cdot (h(i) - h(i-1))$$

If $p = 1/2$ this says that $h(i+1) - h(i) = h(i) - h(i-1)$. In words this says that h has constant slope. Since $h(0) = 0$ and $h(N) = 1$ the slope must be $1/N$ and we must have $h(x) = x/N$. To argue this algebraically, we can instead observe that if $h(i) - h(i-1) = c$ for $1 \leq i \leq N$ then

$$1 = h(N) - h(0) = \sum_{i=1}^{N} h(i) - h(i-1) = Nc$$

so $c = 1/N$. Using the last identity again with the fact that $h(0) = 0$, we have

$$h(x) = h(x) - h(0) = \sum_{i=1}^{x} h(i) - h(i-1) = x/N$$

Recalling the definition of $h(x)$ this means

(6.1) $$P_x(V_N < V_0) = x/N \quad \text{for } 0 \le x \le N$$

To see what the last formula says we will consider a concrete example.

Example 6.2. Matching pennies. Bob, who has 15 pennies, and Charlie, who has 10 pennies, decide to play a game. They each flip a coin. If the two coins match, Bob gets the two pennies (for a profit of 1). If the two coins are different, then Charlie gets the two pennies. They quit when someone has all of the pennies. What is the probability Bob will win the game?

Solution. Let X_n be the number of pennies Bob has after n plays. X_n is a gambler's ruin chain with $p = 1/2$, $N = 25$, and $X_0 = 15$, so by (6.1) the probability Bob wins is 15/25. Notice that the answer is simply Bob's fraction of the total supply of pennies. □

When $p \ne 1/2$ the ideas are the same but the details are more difficult. If we set $c = h(1) - h(0)$ then $(\star)$ implies that for $i \ge 1$

$$h(i) - h(i-1) = c \left(\frac{1-p}{p}\right)^{i-1}$$

Summing from $i = 1$ to N, we have

$$1 = h(N) - h(0) = \sum_{i=1}^{N} h(i) - h(i-1) = c \sum_{i=1}^{N} \left(\frac{1-p}{p}\right)^{i-1}$$

Recalling that for $\theta \ne 1$ the partial sum of the geometric series is

$$\sum_{j=0}^{N-1} \theta^j = \frac{1-\theta^N}{1-\theta}$$

we see that $c = (1-\theta)/(1-\theta^N)$ with $\theta = (1-p)/p$. Summing and using the fact that $h(0) = 0$, we have

$$h(x) = h(x) - h(0) = c \sum_{i=1}^{x} \theta^{i-1} = c \cdot \frac{1-\theta^x}{1-\theta} = \frac{1-\theta^x}{1-\theta^N}$$

Recalling the definition of $h(x)$ and rearranging the fraction we have

(6.2) $$P_x(V_N < V_0) = \frac{\theta^x - 1}{\theta^N - 1} \quad \text{where } \theta = (1-p)/p$$

68   Chapter 1   Markov Chains

To see what (6.2) says we consider:

Example 6.3. Roulette. If we bet $1 on red on a roulette wheel with 18 red, 18 black, and 2 green (0 and 00) holes, we win $1 with probability $18/38 = .4737$ and lose $1 with probability $20/38$. Suppose we bring $50 to the casino with the hope of reaching $100 before going bankrupt. What is the probability we will succeed?

Solution. Here $\theta = (1-p)/p = 20/18$, so (6.2) implies

$$P_{50}(S_T = 100) = \frac{\left(\frac{20}{18}\right)^{50} - 1}{\left(\frac{20}{18}\right)^{100} - 1}$$

Using $(20/18)^{50} = 194$, we have

$$P_{50}(S_T = 100) = \frac{194 - 1}{(194)^2 - 1} = \frac{1}{194 + 1} = .005128 \qquad \square$$

In the gambler's ruin example above we showed by hand that (6.1) had only one solution. In some cases we will want to guess and verify the answer. In those situations it is nice to know that the solution is unique.

(6.3) Lemma. *Consider a Markov chain with finite state space S. Let a and b be two points in S, and let $C = S - \{a, b\}$. Suppose $h(a) = 1$, $h(b) = 0$, and that for $x \in C$ we have*

$$h(x) = \sum_y p(x, y) h(y)$$

If $P_x(V_a \wedge V_b < \infty) > 0$ for all $x \in C$, then $h(x) = P_x(V_a < V_b)$.

Proof. Let $\bar{X}_n$ be the chain modified so that a and b are absorbing states. Our assumptions imply $h(x) = E_x h(\bar{X}_1)$. Iterating and using the Markov property we have $h(x) = E_x h(\bar{X}_n)$. Since the state space is finite and $P_x(V_a \wedge V_b < \infty) > 0$, the pedestrian lemma, (3.3), implies that $\bar{X}_n$ will eventually get stuck at a or b. This means as $n \to \infty$, $E_x h(\bar{X}_n) \to P_x(V_a < V_b)$. Since each term in the sequence is equal to $h(x)$ it follows that $h(x) = P_x(V_a < V_b)$. $\qquad \square$

Using (6.3), we can continue our investigation of Example 1.7.

Example 6.4. Wright–Fisher model with no mutation. The state space is $S = \{0, 1, \ldots N\}$ and the transition probability is

$$p(x, y) = \binom{N}{y} \left(\frac{x}{N}\right)^y \left(\frac{N-x}{N}\right)^{N-y}$$

The right-hand side is the binomial($N, x/N$) distribution, i.e., the number of successes in N trials when success has probability x/N, so the mean number of successes is x. From this it follows that if we define $h(x) = x/N$, then

$$h(x) = \sum_y p(x,y) h(y)$$

Taking $a = N$ and $b = 0$, we have $h(a) = 1$ and $h(b) = 0$. Since $P_x(V_a \wedge V_b < \infty) > 0$ for all $0 < x < N$, it follows from (6.3) that

$$P_x(V_N < V_0) = x/N$$

i.e., the probability of fixation to all A's is equal to the fraction of the genes that are A. □

In addition to figuring where the chain will go, we can also use first-step calculations to compute how long it will take to get where it is going.

Example 6.5. Duration of fair games. Consider the gambler's ruin chain in which $p(i, i+1) = p(i, i-1) = 1/2$ for $0 < i < N$ and the end points are absorbing states: $p(0,0) = 1$ and $p(N, N) = 1$. Let $\tau = \min\{n : X_n \notin (0, N)\}$ be the time at which the chain enters an absorbing state. Let $h(i) = E_i T$ be the expected time to absorption starting from i. Clearly, $h(0) = h(N) = 0$. If $0 < i < N$ then by considering what happens on the first step we have

$$h(i) = 1 + \frac{h(i+1) + h(i-1)}{2}$$

Rearranging, we have

(*) $$h(i+1) - h(i) = -2 + h(i) - h(i-1)$$

Summing from $i = 1$ to $N - 1$, we have

$$h(N) - h(1) = -2(N-1) + h(N-1) - h(0)$$

Now $h(0) = h(N) = 0$ by definition and from symmetry $h(1) = h(N-1)$, so we have $h(1) = (N-1)$. Using (*) it follows that $h(2) - h(1) = -2 + (N-1)$,

$$h(3) - h(2) = -2 + h(2) - h(1) = 4 + (N-1)$$

or in general that

$$h(i+1) - h(i) = -2i + (N-1)$$

Summing from $i = 0$ to $x - 1$ and recalling $\sum_{i=1}^{k} i = k(k+1)/2$ we have

$$h(x) = -x(x-1) + x(N-1) = x(N-x)$$

Recalling the definition of $h(x)$ we have

(6.4) $$E_x \tau = x(N-x)$$

To see what formula (6.4) says, consider Example 6.2. There $N = 25$ and $x = 15$, so the game will take $15 \cdot 10 = 150$ flips on the average. If there are twice as many coins, $N = 50$ and $x = 30$, then the game takes $30 \cdot 20 = 600$ flips on the average, or four times as long.

EXERCISE 6.1. When $p \neq q$ considering what happens on the first step gives $h(i) = 1 + ph(i+1) + qh(i-1)$. Check that the solution of this recursion with $h(0) = h(N) = 0$ is

$$h(i) = \frac{i}{q-p} - \frac{N}{q-p} \cdot \frac{1-(q/p)^i}{1-(q/p)^N}$$

Example 6.6. Waiting time for TT. Let N be the (random) number of times we need to flip a coin before we have gotten Tails on two consecutive tosses. To compute the expected value of N we will introduce a Markov chain with states 0, 1, 2 = the number of Tails we have in a row.

Since getting a Tails increases the number of Tails we have by 1, but getting a Heads sets the number we have to 0, the transition matrix is

	0	1	2
0	1/2	1/2	0
1	1/2	0	1/2
2	0	0	1

Since we are not interested in what happens after we reach 2 we have made 2 an absorbing state. If we let $h(x) = E_x T_2$ then one step reasoning gives

$$h(0) = 1 + .5h(0) + .5h(1), \qquad h(1) = 1 + .5h(0)$$

Here, the first equation says that if we start from 0 and take 1 step then 1/2 of the time we will be back at 0 and 1/2 of the time we will be at 1. The same reasoning applies to the second equation. It looks different since $h(2) = 0$ and disappears. Plugging the second equation into the first gives $h(0) = 1.5 + .75h(0)$, so $.25h(0) = 1.5$ or $h(0) = 6$. □

A second method of computing $E_y T_y$ comes from the strong law in (4.9):

(6.5) Theorem. If p is an irreducible transition probability and has stationary distribution π then
$$\pi(y) = \frac{1}{E_y T_y}$$

Why is this true? Recall that if $N_n(y)$ is the number of visits to y by time n, then (4.9) says $N_n(y)/n \to \pi(y)$. On the other hand, the times between returns to y are independent with mean $E_y T_y$, so it follows from the strong law of large numbers that in the long run we will make one visit to y every $E_y T_y$ units of time, i.e., $N_n(y)/n \to 1/E_y T_y$. □

The proof of (6.5) will be given in Section 1.8. To illustrate it use consider

Example 6.7. Waiting time for HT. Suppose we start flipping a coin and stop when we first complete the pattern HT, i.e., have a heads followed by a tails. How long do we expect this to take?

Solution 1. Consider the Markov chain X_n that gives the outcome of the last two flips. X_n is Markov chain with transition probability:

	HH	HT	TH	TT
HH	1/2	1/2	0	0
HT	0	0	1/2	1/2
TH	1/2	1/2	0	0
TT	0	0	1/2	1/2

Since this matrix is doubly stochastic, the stationary distribution assigns probability 1/4 to each state. One can verify this and check that convergence to equilibrium is rapid by noting that all the entries of p^2 are equal to 1/4.

Using (6.5) now, we have
$$E(T_{HT}|X_0 = HT) = \frac{1}{\pi(HT)} = 4$$

To get from this to what we wanted to calculate, note that if we start with a H at time -1 and a T at time 0, then we have nothing that will help us in the future, so the expected waiting time for a HT when we start with nothing is the same.

Solution 2. For a simple derivation of $ET_{HT} = 4$, consider the possible realization T, T, H, H, H, T and note that to get HT we first have to wait for a H, then the next T completes the pattern. The number of steps to get the first H is geometric with success probability $p = 1/2$ and hence mean 2. Once we get the first heads the expected waiting time for the tails also has mean 2, so the total expected waiting time is 4.

Back to TT. At this point we have seen that the expected waiting time for HT is 4, but the expected waiting time for TT is 6. This difference can be explained by deriving the answer for TT by using the method in Solution 1. To do this, we begin by using (6.5) to conclude

$$E(T_{TT}|X_0 = TT) = \frac{1}{\pi(TT)} = 4$$

However this time if we start with a T at time -1 and a T at time 0, so a T at time 1 will give us a TT and a return at time 1, while if we get a H at time 1 we have wasted 1 turn and we have nothing that can help us later, so if EN_{TT} is the expected waiting time for TT when we start with nothing, then

$$4 = E(T_{TT}|X_0 = TT) = \frac{1}{2} \cdot 1 + \frac{1}{2} \cdot (1 + EN_{TT})$$

Solving gives $EN_{TT} = 6$, so it takes longer to observe TT. The reason for this, which can be seen in the last equation, is that once we have one TT, we will get another one with probability 1/2, while occurrences of HT cannot overlap.

Waiting times for three coin patterns. Let N_{xyz} be the number of tosses needed to observe the coin pattern xyz. By symmetry the waiting time for a pattern and its opposite with heads and tails interchanged are the same, e.g., $EN_{HHT} = EN_{TTH}$. Thus without loss of generality we can suppose the first coin is H and reduce the number of cases to be considered to 4.

EXERCISE 6.2. Show that $EN_{HTT} = 8$ and $EN_{HHT} = 8$.

EXERCISE 6.3. Use the method of Example 6.6 to show $EN_{HHH} = 14$.

EXERCISE 6.4. Argue that $EN_{HTH} = 2(1 + EN_{HT})$ and hence $EN_{HTH} = 10$.

Example 6.8. A sucker bet. Consider the following gambling game. Player 1 picks a three-coin pattern, then Player 2 picks a different three-coin pattern. One coin is flipped repeatedly. The person whose pattern appears first wins a dollar from the other. Somewhat surprisingly Player 2 has a considerable advantage in this game. Suppose without loss of generality that Player 1 picks a three-coin pattern that begins with H.

Player 1	Player 2	Prob. 2 wins
HHH	THH	$7/8$
HHT	THH	$3/4$
HTH	HHT	$2/3$
HTT	HHT	$2/3$

To check the first line, note that Player 1 wins if the first three tosses are HHH but loses otherwise, since the occurrence of the first HH after the first T leads to a victory for player 2. To help see this consider the following example

$$HHTHTTHTHH$$

The second line is similar, Player 1 wins if the first two tosses are HH, but always loses otherwise.

To check the third line, note that nothing can happen until the first heads appears. If the next toss is H, then Player 2 is certain to win. If the next two tosses are TH, then Player 1 wins. If the next two tosses are TT then the game starts over. This implies that if p is Player 2's probability of winning,

$$p = 1/2 + (1/4)p \quad \text{so} \quad (3/4)p = 2/4$$

The fourth line is almost the same, but now the third sentence changes to: "If the next two tosses are TT, then Player 1 wins. If the next two tosses are TH, then the game starts over."

1.7. Infinite State Spaces

In this section we delve into the complications that can arise when the state space for the chain is infinite. Our discussions will focus primarily on two examples.

Example 7.1. Reflecting random walk. Imagine a particle that moves on $\{0, 1, 2, \ldots\}$ according to the following rules. It takes a step to the right with probability p. It attempts to take a step to the left with probability $1 - p$, but if it is at 0 and tries to jump to the left, it stays at 0, since there is no -1 to jump to. In symbols,

$$p(i, i+1) = p \quad \text{when } i \geq 0$$
$$p(i, i-1) = 1 - p \quad \text{when } i \geq 1$$
$$p(0, 0) = 1 - p$$

This is a birth and death chain, so we can solve for the stationary distribution using the detailed balance equations:

$$p\pi(i) = (1-p)\pi(i+1) \quad \text{when } i \geq 0$$

74 Chapter 1 Markov Chains

Rewriting this as $\pi(i+1) = \pi(i) \cdot p/(1-p)$ and setting $\pi(0) = c$, we have

(*) $$\pi(i) = c\left(\frac{p}{1-p}\right)^i$$

There are now three cases to consider:

$p > 1/2$: $p/(1-p) > 1$, so $\pi(i)$ increases exponentially fast and π cannot possibly be a stationary distribution.

$p = 1/2$: $p/(1-p) = 1$, so $\pi(i) = c$ and $\sum_i \pi(i) = \infty$. Again, there is no stationary distribution.

$p < 1/2$: $p/(1-p) < 1$. $\pi(i)$ decreases exponentially fast, so $\sum_i \pi(i) < \infty$, and we can pick c to make π a stationary distribution.

To find the value of c to make π a probability distribution we recall $\sum_{i=0}^{\infty} \theta^i = 1/(1-\theta)$ when $\theta < 1$. Taking $\theta = p/(1-p)$ and hence $1 - \theta = (1-2p)/(1-p)$, we see that the sum of the $\pi(i)$ defined in (*) is $c(1-p)/(1-2p)$, so

(7.1) $$\pi(i) = \frac{1-2p}{1-p} \cdot \left(\frac{p}{1-p}\right)^i = (1-\theta)\theta^i$$

To confirm that we have succeeded in making the $\pi(i)$ add up to 1, note that if we are flipping a coin with a probability θ of Heads, then the probability of getting i Heads before we get our first Tails is given by $\pi(i)$.

The reflecting random walk is clearly irreducible. To check that it is aperiodic note that $p(0,0) > 0$ implies 0 has period 1, and then (4.3) implies that all states have period 1. Using the convergence theorem, (4.5), now we see that

I. When $p < 1/2$, $P(X_n = j) \to \pi(j)$, the stationary distribution in (7.1).

It should not be surprising that the system stabilizes when $p < 1/2$. In this case movements to the left have a higher probability than to the right, so if the chain gets to a large value, then it drifts back toward 0. On the other hand if steps to the right are more frequent than those to the left, then the chain will drift to the right and wander off to ∞.

II. When $p > 1/2$, $X_n \to \infty$ as $n \to \infty$, so all states are transient.

Why is this true? Because of the reflection at 0, "$X_n - X_0$ is larger than $S_n = x_1 + \cdots + x_n$ where $x_1, x_2, \ldots$ are independent and have $P(x_i = 1) = p$ and $P(x_i = -1) = 1 - p$." The law of large numbers implies

$$S_n/n \to Ex_i = p(1) + (1-p)(-1) = 2p - 1 > 0$$

so $S_n \to \infty$ and it follows that $X_n \to \infty$. □

Proof. To turn the idea above into a proof we have to make the phrase in quotes precise. Let $x_m = 1$ if the mth step was to the right and $x_m = -1$ if the mth step was to the left (counting a 0 to 0 transition as a step to left). It is clear from the definition that $X_m - X_{m-1} \geq x_m$. Summing for $m = 1, 2, \ldots, n$ we have $X_n - X_0 \geq x_1 + x_2 + \cdots + x_n$. □

To figure out what happens in the borderline case $p = 1/2$, we use results from Section 1.6. Recall we have defined $V_y = \min\{n \geq 0 : X_n = y\}$ and (6.1) tells us that if $x > 0$

$$P_x(V_N < V_0) = x/N$$

If we keep x fixed and let $N \to \infty$, then $P_x(V_N < V_0) \to 0$ and hence

$$P_x(V_0 < \infty) = 1$$

In words, for any starting point x, the random walk will return to 0 with probability 1. To compute the mean return time we note that if $\tau_N = \min\{n : X_n \notin (0, N)\}$, then we have $\tau_N \leq V_0$ and by (6.4) we have $E_1 \tau_N = N - 1$. Letting $N \to \infty$ and combining the last two facts shows $E_1 V_0 = \infty$. Reintroducing our old hitting time $T_0 = \min\{n > 0 : X_n = 0\}$ and noting that on our first step we go to 0 or to 1 with probability 1/2 shows that

$$E_0 T_0 = (1/2) \cdot 1 + (1/2) E_1 V_0 = \infty$$

Summarizing the last two paragraphs, we have

III. When $p = 1/2$, $P_0(T_0 < \infty) = 1$ but $E_0 T_0 = \infty$.

Thus when $p = 1/2$, 0 is recurrent in the sense we will certainly return, but it is not recurrent in the following sense:

x is said to be **positive recurrent** if $E_x T_x < \infty$.

If a state is recurrent but not positive recurrent, i.e., $P_x(T_x < \infty) = 1$ but $E_x T_x = \infty$, then we say that x is **null recurrent**.

In our new terminology, our results for reflecting random walk say

If $p < 1/2$, 0 is positive recurrent

If $p = 1/2$, 0 is null recurrent

If $p > 1/2$, 0 is transient

In reflecting random walk, null recurrence thus represents the borderline between recurrence and transience. This is what we think in general when we hear the term. To see the reason we might be interested in positive recurrence recall that by (6.5)

$$\pi(x) = \frac{1}{E_x T_x}$$

If $E_x T_x = \infty$, then this gives $\pi(x) = 0$. With a lot of imagination the last observation can be turned into the following:

(7.2) Theorem. *For an irreducible chain the following are equivalent:*
   (i) Some state is positive recurrent.
   (ii) All states are positive recurrent.
   (iii) There is a stationary distribution π.

The proof of (7.2) is again postponed to Section 1.8, and, for the moment, we content ourselves to understand what this says. The equivalence of (i) and (ii) is another expression of solidarity. Their equivalence to (iii) is consistent with our findings for reflecting random walk. When $p \geq 1/2$, 0 is transient or null recurrent, and there is no stationary distribution. When $p < 1/2$, there is a stationary distribution and (6.5) implies

(7.3) $$E_0 T_0 = \frac{1}{\pi(0)} = \frac{1}{1-\theta} = \frac{1-p}{1-2p}$$

so 0 is positive recurrent.

EXERCISE 7.1. (a) Use gambling to argue that if $p < 1/2$, $E_1 T_0 = 1/(1-2p)$. (b) Argue that $E_0(T_0 - 1) = p E_1 T_0$ and use this to get another derivation of (7.3).

The notions of positive and null recurrence are somewhat subtle, and occasionally require considerable ingenuity to check. We mention these concepts primarily to make contact with the other way in which the basic convergence theorem can be stated. In most of our examples, we do not check positive recurrence but instead will solve $\pi p = \pi$ and then apply (4.4).

(7.4) Convergence theorem, II. *Suppose p is irreducible, aperiodic, and all states are positive recurrent. Then*

$$\lim_{n \to \infty} p^n(x,y) = \pi(y)$$

where π is the unique nonnegative solution of $\pi p = \pi$ with $\sum_i \pi(i) = 1$.

Proof. This follows easily from our other results. By (7.2) the assumption that all states are positive recurrent implies that there is a stationary distribution. Using (4.5) now it follows that $\lim_{n\to\infty} p^n(x,y) = \pi(y)$. The uniqueness of the stationary distribution claimed in the last sentence follows from (4.6). □

Our next example may at first seem to be quite different. In a branching process 0 is an absorbing state, so by (3.2) all the other states are transient. However, as the story unfolds we will see that branching processes have the same trichotomy as random walks do.

Example 7.2. Branching Processes. Consider a population in which each individual in the nth generation gives birth to an independent and identically distributed number of children. The number of individuals at time n, X_n is a Markov chain with transition probability given in Example 1.6. As announced there were are interested in the question:

Q. What is the probability the species avoids extinction?

where "extinction" means becoming absorbed state at 0. As we will now explain, whether this is possible or not can be determined by looking at the average number of offspring of one individual:

$$\mu = \sum_{k=0}^{\infty} k p_k$$

If there are m individuals at time $n - 1$, then the mean number at time n is $m\mu$. More formally the conditional expectation given X_{n-1}

$$E(X_n | X_{n-1}) = \mu X_{n-1}$$

Taking expected values of both sides gives $EX_n = \mu EX_{n-1}$. Iterating gives

(7.5) $$EX_n = \mu^n EX_0$$

If $\mu < 1$, then $EX_n \to 0$ exponentially fast. Using the inequality

$$EX_n \geq P(X_n \geq 1)$$

it follows that $P(X_n \geq 1) \to 0$ and we have

I. *If $\mu < 1$ then extinction occurs with probability 1.*

To bring out the parallel with conclusion I for the random walk do

78 Chapter 1 Markov Chains

EXERCISE 7.2. Modify the branching process so that $p(0,1) = 1$. In words, whenever the population dies out we start it with one individual in the next time period. Show that if $\mu < 1$ then the modified branching process is positive recurrent.

To treat the cases $\mu \geq 1$ we will use a one-step calculation. Let ρ be the probability that this process dies out (i.e., reaches the absorbing state 0) starting from $X_0 = 1$. If there are k children in the first generation, then in order for extinction to occur, the family line of each child must die out, an event of probability ρ^k, so we can reason that

($\oplus$)
$$\rho = \sum_{k=0}^{\infty} p_k \rho^k$$

If we let $\varphi(\theta) = \sum_{k=0}^{\infty} p_k \theta^k$ be the generating function of the distribution p_k, then the last equation can be written simply as $\rho = \varphi(\rho)$.

The equation in ($\oplus$) has a trivial root at $\rho = 1$ since $\varphi(\rho) = \sum_{k=0}^{\infty} p_k = 1$. The next result identifies the root that we want:

(7.6) Lemma. *The extinction probability ρ is the smallest solution of the equation $\varphi(x) = x$ with $0 \leq x \leq 1$.*

Here a picture is worth a hundred words.

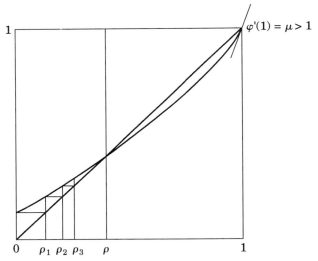

Proof. Extending the reasoning for ($\oplus$) we see that in order for the process to hit 0 by time n, all of the processes started by first-generation individuals must

hit 0 by time $n-1$, so

$$P(X_n = 0) = \sum_{k=0}^{\infty} p_k P(X_{n-1} = 0)^k$$

From this we see that if $\rho_n = P(X_n = 0)$ for $n \geq 0$, then $\rho_n = \varphi(\rho_{n-1})$ for $n \geq 1$.

Since 0 is an absorbing state, $\rho_0 \leq \rho_1 \leq \rho_2 \leq \ldots$ and the sequence converges to a limit ρ_∞. Letting $n \to \infty$ in $\rho_n = \varphi(\rho_{n-1})$ implies that $\rho_\infty = \varphi(\rho_\infty)$, i.e., ρ_∞ is a solution of $\varphi(x) = x$. To complete the proof now let ρ be the smallest solution. Clearly $\rho_0 = 0 \leq \rho$. Using the fact that φ is increasing, it follows that $\rho_1 = \varphi(\rho_0) \leq \varphi(\rho) = \rho$. Repeating the argument we have $\rho_2 \leq \rho$, $\rho_3 \leq \rho$ and so on. Taking limits we have $\rho_\infty \leq \rho$. However, ρ is the smallest solution, so we must have $\rho_\infty = \rho$. □

To see what this says, let us consider a concrete example.

Example 7.3. Binary branching. Suppose $p_2 = a$, $p_0 = 1-a$, and the other $p_k = 0$. In this case $\varphi(\theta) = a\theta^2 + 1 - a$, so $\varphi(x) = x$ means

$$0 = ax^2 - x + 1 - a = (x-1)(ax - (1-a))$$

The roots are 1 and $(1-a)/a$. If $a \leq 1/2$, then the smallest root is 1, while if $a > 1/2$ the smallest root is $(1-a)/a$.

Noting that $a \leq 1/2$ corresponds to mean $\mu \leq 1$ in binary branching motivates the following guess:

II. If $\mu > 1$, then there is positive probability of avoiding extinction.

Proof. In view of (7.6) we only have to show there is a root < 1. We begin by discarding a trivial case. If $p_0 = 0$, then $\varphi(0) = 0$, 0 is the smallest root, and there is no probability of dying out. If $p_0 > 0$, then $\varphi(0) = p_0 > 0$. Differentiating the definition of φ, we have

$$\varphi'(x) = \sum_{k=1}^{\infty} p_k \cdot kx^{k-1} \quad \text{so} \quad \varphi'(1) = \sum_{k=1}^{\infty} kp_k = \mu$$

If $\mu > 1$ then the slope of φ at $x = 1$ is larger than 1, so if ϵ is small, then $\varphi(1-\epsilon) < 1-\epsilon$. Combining this with $\varphi(0) > 0$ we see there must be a solution of $\varphi(x) = x$ between 0 and $1-\epsilon$. See the figure in the proof of (7.6). □

Turning to the borderline case:

III. *If $\mu = 1$ and we exclude the trivial case $p_1 = 1$, then extinction occurs with probability 1.*

Proof. By (7.6) we only have to show that there is no root < 1. To do this we note that if $p_1 < 1$, then for $y < 1$

$$\varphi'(y) = \sum_{k=1}^{\infty} p_k \cdot kx^{k-1} < \sum_{k=1}^{\infty} p_k k = 1$$

so if $x < 1$ then $\varphi(x) = \varphi(1) - \int_x^1 \varphi'(y)\, dy > 1 - (1-x) = x$. Thus $\varphi(x) > x$ for all $x < 1$. □

Note that in binary branching with $a = 1/2$, $\varphi(x) = (1+x^2)/2$, so if we try to solve $\varphi(x) = x$ we get

$$0 = 1 - 2x + x^2 = (1-x)^2$$

i.e., a double root at $x = 1$. In general when $\mu = 1$, the graph of φ is tangent to the diagonal (x, x) at $x = 1$. This slows down the convergence of ρ_n to 1 so that it no longer occurs exponentially fast.

In more advanced treatments, it is shown that if the offspring distribution has mean 1 and variance $\sigma^2 > 0$, then

$$P_1(X_n > 0) \sim \frac{2}{n\sigma^2}$$

This is not easy even for the case of binary branching, so we refer to reader to Section 1.9 of Athreya and Ney (1972) for a proof. We mention the result here because it allows us to see that the expected time for the process to die out $\sum_n P_1(T_0 > n) = \infty$. As in Exercise 7.2, if we modify the branching process, so that $p(0, 1) = 1$ then in the modified process

If $\mu < 1$, 0 is positive recurrent

If $\mu = 1$, 0 is null recurrent

If $\mu > 1$, 0 is transient

Our final example gives an application of branching processes to queueing theory.

Example 7.4. M/G/1 queue. We will not be able to explain the name of this example until we consider continuous-time Markov chains in Chapter

4. However, imagine a queue of people waiting to use an automated teller machine. Let X_n denote the number of people in line at the moment of the departure of the nth customer. To model this as a Markov chain we let a_k be the probability k customers arrive during one service time and write down the transition probability

$$p(0, k) = a_k \quad \text{and} \quad p(i, i - 1 + k) = a_k \quad \text{for } k \geq 0$$

with $p(i, j) = 0$ otherwise.

To explain this, note that if there is a queue, it is reduced by 1 by the departure of a customer, but k new customers will come with probability k. On the other hand if there is no queue, we must first wait for a customer to come and the queue that remains at her departure is the number of customers that arrived during her service time. The pattern becomes clear if we write out a few rows and columns of the matrix:

	0	1	2	3	4	5	...
0	a_0	a_1	a_2	a_3	a_4	a_5	
1	a_0	a_1	a_2	a_3	a_4	a_5	
2	0	a_0	a_1	a_2	a_3	a_4	
3	0	0	a_0	a_1	a_2	a_3	
4	0	0	0	a_0	a_1	a_2	

If we regard the customers that arrive during a person's service time to be her children, then this queueing process gives rise to a branching process. From the results above for branching processes we see that if we denote the mean number of children by $\mu = \sum_k k a_k$, then

If $\mu < 1$, 0 is positive recurrent

If $\mu = 1$, 0 is null recurrent

If $\mu > 1$, 0 is transient

To bring out the parallels between the three examples, note that when $\mu > 1$ or $p > 1/2$ the process drifts away from 0 and is transient. When $\mu < 1$ or $p < 1/2$ the process drifts toward 0 and is positive recurrent. When $\mu = 1$ or $p = 1/2$, there is no drift. The process eventually hits 0 but not in finite expected time, so 0 is null recurrent.

1.8. Proofs of the Convergence Theorems

At this point we owe the reader the proofs of (4.5), (4.7), (4.8), (6.5), and (7.2). To prepare for the proof of the convergence theorem (4.5), we need the following:

82   Chapter 1   Markov Chains

(8.1) **Lemma.** *If there is a stationary distribution, then all states y that have $\pi(y) > 0$ are recurrent.*

Proof. (3.11) tells us that $E_x N(y) = \sum_{n=1}^{\infty} p^n(x,y)$, so

$$\sum_x \pi(x) E_x N(y) = \sum_x \pi(x) \sum_{n=1}^{\infty} p^n(x,y)$$

Interchanging the order of summation and using $\pi p^n = \pi$, the above

$$= \sum_{n=1}^{\infty} \sum_x \pi(x) p^n(x,y) = \sum_{n=1}^{\infty} \pi(y) = \infty$$

since $\pi(y) > 0$. Using (3.8) now gives $E_x N(y) = \rho_{xy}/(1 - \rho_{yy})$, so

$$\infty = \sum_x \pi(x) \frac{\rho_{xy}}{1 - \rho_{yy}} \le \frac{1}{1 - \rho_{yy}}$$

the second inequality following from the facts that $\rho_{xy} \le 1$ and π is a probability measure. This shows that $\rho_{yy} = 1$, i.e., y is recurrent. □

With (8.1) in hand we are ready to tackle the proof of:

(4.5) **Convergence theorem.** *Suppose p is irreducible, aperiodic, and has stationary distribution π. Then as $n \to \infty$, $p^n(x,y) \to \pi(y)$.*

Proof. Let $S^2 = S \times S$. Define a transition probability $\bar{p}$ on $S \times S$ by

$$\bar{p}((x_1, y_1), (x_2, y_2)) = p(x_1, x_2) p(y_1, y_2)$$

In words, each coordinate moves independently. Our first step is to check that $\bar{p}$ is irreducible. This may seem like a silly thing to do first, but this is the only step that requires aperiodicity. Since p is irreducible, there are K, L, so that $p^K(x_1, x_2) > 0$ and $p^L(y_1, y_2) > 0$. Since x_2 and y_2 have period 1, it follows from (4.2) that if M is large, then $p^{L+M}(x_2, x_2) > 0$ and $p^{K+M}(y_2, y_2) > 0$, so

$$\bar{p}^{K+L+M}((x_1, y_1), (x_2, y_2)) > 0$$

Our second step is to observe that since the two coordinates are independent $\bar{\pi}(a,b) = \pi(a)\pi(b)$ defines a stationary distribution for $\bar{p}$, and (8.1) implies that all states are recurrent for $\bar{p}$. Let (X_n, Y_n) denote the chain on $S \times S$, and let T be the first time that the two coordinates are equal, i.e.,

Section 1.8 Proofs of the Convergence Theorems 83

$T = \min\{n \geq 0 : X_n = Y_n\}$. Let $V_{(x,x)} = \min\{n \geq 0 : X_n = Y_n = x\}$ be the time of the first visit to (x,x). Since $\bar{p}$ is irreducible and recurrent, $V_{(x,x)} < \infty$ with probability one. Since $T \leq V_{(x,x)}$ for our favorite x we must have $T < \infty$.

The third and somewhat magical step is to prove that on $\{T \leq n\}$, the two coordinates X_n and Y_n have the same distribution. By considering the time and place of the first intersection and then using the Markov property we have

$$P(X_n = y, T \leq n) = \sum_{m=1}^{n} \sum_{x} P(T = m, X_m = x, X_n = y)$$
$$= \sum_{m=1}^{n} \sum_{x} P(T = m, X_m = x) P(X_n = y | X_m = x)$$
$$= \sum_{m=1}^{n} \sum_{x} P(T = m, Y_m = x) P(Y_n = y | Y_m = x)$$
$$= P(Y_n = y, T \leq n)$$

To finish up we observe that using the last equality we have

$$P(X_n = y) = P(X_n = y, T \leq n) + P(X_n = y, T > n)$$
$$= P(Y_n = y, T \leq n) + P(X_n = y, T > n)$$
$$\leq P(Y_n = y) + P(X_n = y, T > n)$$

and similarly $P(Y_n = y) \leq P(X_n = y) + P(Y_n = y, T > n)$. So

$$|P(X_n = y) - P(Y_n = y)| \leq P(X_n = y, T > n) + P(Y_n = y, T > n)$$

and summing over y gives

$$\sum_{y} |P(X_n = y) - P(Y_n = y)| \leq 2P(T > n)$$

If we let $X_0 = x$ and let Y_0 have the stationary distribution π, then Y_n has distribution π, and it follows that

$$\sum_{y} |p^n(x, y) - \pi(y)| \leq 2P(T > n) \to 0$$

proving the desired result. □

Next on our list is the equivalence of positive recurrence and the existence of a stationary distribution, (7.2), the first piece of which is:

(8.2) Theorem. Let x be a positive recurrent state, let $T_x = \inf\{n \geq 1 : X_n = x\}$, and let

$$\mu(y) = \sum_{n=0}^{\infty} P_x(X_n = y, T_x > n)$$

Then $\pi(y) = \mu(y)/E_x T_x$ defines a stationary distribution.

To prepare for the proof of (6.5) note that $\mu(x) = 1$ so $\pi(x) = 1/E_x T_x$. Another useful bit of trivia that explains the norming constant is that the definition and (3.10) imply

$$\sum_{y \in S} \mu(y) = \sum_{n=0}^{\infty} P_x(T_x > n) = E_x T_x$$

Why is this true? This is called the "cycle trick." $\mu(y)$ is the expected number of visits to y in $\{0, \ldots, T_x - 1\}$. Multiplying by p moves us forward one unit in time so $\mu p(y)$ is the expected number of visits to y in $\{1, \ldots, T_x\}$. Since $X(T_x) = X_0 = x$ it follows that $\mu = \mu p$. Since π is just μ divided a constant to make the sum 1, π is a stationary distribution.

Proof. To formalize this intuition, let $\bar{p}_n(x, y) = P_x(X_n = y, T_x > n)$ and interchange sums to get

$$\sum_y \mu(y) p(y, z) = \sum_{n=0}^{\infty} \sum_y \bar{p}_n(x, y) p(y, z)$$

Case 1. Consider the generic case first: $z \neq x$.

$$\sum_y \bar{p}_n(x, y) p(y, z) = \sum_y P_x(X_n = y, T_x > n, X_{n+1} = z)$$
$$= P_x(T_x > n+1, X_{n+1} = z) = \bar{p}_{n+1}(x, z)$$

Here the second equality holds since the chain must be somewhere at time n, and the third is just the definition of $\bar{p}_{n+1}$. Summing from $n = 0$ to ∞, we have

$$\sum_{n=0}^{\infty} \sum_y \bar{p}_n(x, y) p(y, z) = \sum_{n=0}^{\infty} \bar{p}_{n+1}(x, z) = \mu(z)$$

since $\bar{p}_0(x, z) = 0$.

Case 2. Now suppose that $z = x$. Reasoning as above we have

$$\sum_y \bar{p}_n(x, y) p(y, x) = \sum_y P_x(X_n = y, T_x > n, X_{n+1} = x) = P_x(T_x = n+1)$$

Summing from $n = 0$ to ∞ we have

$$\sum_{n=0}^{\infty} \sum_y \bar{p}_n(x,y) p(y,x) = \sum_{n=0}^{\infty} P_x(T_x = n+1) = 1 = \mu(x)$$

since $P_x(T = 0) = 0$. □

With (8.2) established we can now easily prove:

(4.7) Theorem. *If the state space S is finite then there is at least one stationary distribution.*

Proof. By (3.5) we can restrict our attention to a closed irreducible subset of S, and hence suppose without loss of generality that the chain is irreducible. Let $y \in S$. In view of (8.2) it is enough to prove that y is positive recurrent, i.e., $E_y T_y < \infty$. To do this we note that irreducibility implies that for each x there is a $k(x)$ so that $P_x(T_y \leq k(x)) > 0$. Since S is finite, $K = \max\{k(x) : x \in S\} < \infty$, and there is an $\alpha > 0$ so that $P_x(T_y \leq K) \geq \alpha$. The pedestrian lemma (3.3) now implies that $P_x(T_y > nK) \leq (1-\alpha)^n$, so $E_x T_y < \infty$ for all $x \in S$ and in particular $E_y T_y < \infty$. □

To prepare for the second piece of (7.2) we now prove:

(8.3) Theorem. *Suppose p is irreducible. Then for any $x \in S$, as $n \to \infty$*

$$\frac{N_n(y)}{n} \to \frac{1}{E_y T_y}$$

Proof. Consider the first the case in which y is transient. (3.10) implies that $EN_n(y) < \infty$ so $N_n(y) < \infty$ and hence $N_n(y)/n \to 0$ as $n \to \infty$. On the other hand transience implies $P_y(T_y = \infty) > 0$, so $E_y T_y = \infty$ and $1/E_y T_y = 0$.

Turning to the recurrent case, suppose that we start at y. Let $R(k) = \min\{n \geq 1 : N_n(y) = k\}$ be the time of the kth return to y. Let $R(0) = 0$ and for $k \geq 1$ let $t_k = R(k) - R(k-1)$. Since we have assumed $X_0 = y$, the times between returns, $t_1, t_2, \ldots$ are independent and identically distributed so the strong law of large numbers for nonnegative random variables implies that

$$R(k)/k \to E_y T_y \leq \infty$$

From the definition of $R(k)$ it follows that $R(N_n(y)) \leq n < R(N_n(y) + 1)$. Dividing everything by $N_n(y)$ and then multiplying and dividing on the end by $N_n(y) + 1$, we have

$$\frac{R(N_n(y))}{N_n(y)} \leq \frac{n}{N_n(y)} < \frac{R(N_n(y)+1)}{N_n(y)+1} \cdot \frac{N_n(y)+1}{N_n(y)}$$

Letting $n \to \infty$, we have $n/N_n(y)$ trapped between two things that converge to $E_y T_y$, so
$$\frac{n}{N_n(y)} \to E_y T_y$$

To generalize now to $x \ne y$, observe that the strong Markov property implies that conditional on $\{T_y < \infty\}$, $t_2, t_3, \ldots$ are independent and identically distributed and have $P_x(t_k = n) = P_y(T_y = n)$ so
$$R(k)/k = t_1/k + (t_2 + \cdots + t_k)/k \to 0 + E_y T_y$$

and we have the conclusion in general. □

From (8.3) we can easily get:

(6.5) Theorem. *If p is an irreducible transition probability and has stationary distribution π, then*
$$\pi(y) = 1/E_y T_y$$

Why is this true? From (8.3) it follows that
$$\frac{N_n(y)}{n} \to \frac{1}{E_y T_y}$$

Taking expected value and using the fact that $N_n(y) \le n$, it can be shown that this implies
$$\frac{E N_n(y)}{n} \to \frac{1}{E_y T_y}$$

By the reasoning that led to (3.11), we have $E_x N_n(y) = \sum_{m=1}^{n} p^m(x, y)$. The convergence theorem implies $p^n(x, y) \to \pi(y)$, so we have
$$\frac{E_x N_n(y)}{n} \to \pi(y)$$

Comparing the last two results gives the desired conclusion. □

We are now ready to put the pieces together.

(7.2) Theorem. *For an irreducible chain the following are equivalent:*

(i) Some x is positive recurrent.

(ii) There is a stationary distribution.

(iii) All states are positive recurrent.

Section 1.8 Proofs of the Convergence Theorems

Proof (8.2) shows that (i) implies (ii). Noting that irreducibility implies $\pi(y) > 0$ for all y and then using (6.7) shows that (ii) implies (iii). It is trivial that (iii) implies (i). □

We are now ready to pay off our last debt.

(4.7) Strong law. Suppose p is irreducible and has stationary distribution π. Let $r(x)$ be the reward we earn in state x and suppose that $\sum \pi(y)|r(y)| < \infty$. Then as $n \to \infty$ we have

$$\frac{1}{n} \sum_{k=1}^{n} r(X_k) \to \sum_{x} \pi(x) r(x)$$

Proof. Let T_x^k be the time of the kth return to x with $T_x^0 = 0$. The strong Markov property implies that

$$Y_k = r(X(T_x^{k-1} + 1)) + \cdots + r(X(T_x^k)), \quad k \geq 1$$

are independent and identically distributed with $E|Y_k| < \infty$. Let $N_n(x)$ be the number of visits to x by time n. Using (8.3) and the strong law of large numbers we have

$$\frac{N_n(x)}{n} \cdot \frac{1}{N_n(x)} \sum_{m=1}^{N_n(x)} Y_m \to \frac{1}{E_x T_x^1} \cdot EY_1 = \sum r(y)\pi(y)$$

where the last equality follows from the construction in (8.2).

The last detail to confront is the fact that

$$\sum_{m=1}^{N_n(x)} Y_m = \sum_{k=1}^{R(N_n(y))} r(X_k)$$

while what we are really interested in is $\sum_{k=1}^{n} r(X_k)$. In the proof of (8.3) we saw that $R(N_n(y)) \leq n < R(N_{n+1}(y))$, so the difference is at most part of one of the Y_m. To take care of the weird things that might happen in the middle of a sum we look instead at the nonnegative random variables

$$W_k = |r(X(T_x^{k-1} + 1))| + \cdots + |r(X(T_x^k))|, \quad k \geq 1$$

Again the W_k are independent and identically distributed with $EW_k < \infty$. The remaining gap can be filled with the following:

(8.4) **Lemma.** *Suppose $W_1, W_2, \ldots$ are independent and identically distributed nonnegative random variables with $E|W_i| < \infty$ then $W_i/i \to 0$ and hence*

$$\max_{1 \le i \le n+1} W_i/(n+1) \to 0$$

To see why this is enough to finish the proof of (4.7) we note that

$$\left| \sum_{k=1}^{R(N_n(y))} r(X_k) - \sum_{k=1}^{n} r(X_k) \right| \le W_{N_n(y)+1}$$

and $N_n(y) \le n$.

Proof. Let $\epsilon > 0$. Since the tail of a distribution function is decreasing,

$$\sum_{i=1}^{\infty} P(W_i/\epsilon > i) \le \int_0^{\infty} P(W_i/\epsilon > x)\, dx < \infty$$

This shows that we only have $W_i > \epsilon i$ finitely many times. Since ϵ is arbitrary, it follows that $W_i/i \to 0$.

To get the second conclusion from the first we note that for any K we have

$$\max_{1 \le i \le n+1} W_i/(n+1) \le \max_{1 \le i \le K} W_i/(n+1) + \max_{K < i \le n+1} W_i/i$$

Let $\epsilon > 0$. If we pick K large enough, the second term is always $< \epsilon$. For any fixed K the first term tends to 0 as $n \to \infty$. These two facts give the desired conclusion. □

1.9. Exercises

9.1. A fair coin is tossed repeatedly with results $Y_0, Y_1, Y_2, \ldots$ that are 0 or 1 with probability 1/2 each. For $n \ge 1$ let $X_n = Y_n + Y_{n-1}$ be the number of 1's in the $(n-1)$th and nth tosses. Is X_n a Markov chain?

9.2. Five white balls and five black balls are distributed in two urns in such a way that each urn contains five balls. At each step we draw one ball from each urn and exchange them. Let X_n be the number of white balls in the left urn at time n. Compute the transition probability for X_n.

9.3. Suppose that the probability it rains today is 0.3 if neither of the last two days was rainy, but 0.6 if at least one of the last two days was rainy. Let the weather on day n, W_n, be R for rain, or S for sun. W_n is not a Markov chain, but the weather for the last two days $X_n = (W_{n-1}, W_n)$ is a Markov chain with four states $\{RR, RS, SR, SS\}$. (a) Compute its transition probability. (b) Compute the two-step transition probability. (c) What is the probability it will rain on Wednesday given that it did not rain on Sunday or Monday.

9.4. Consider a gambler's ruin chain with $N = 4$. That is, if $1 \leq i \leq 3$, $p(i, i+1) = 0.4$, and $p(i, i-1) = 0.6$, but the endpoints are absorbing states: $p(0,0) = 1$ and $p(4,4) = 1$ Compute $p^3(1,4)$ and $p^3(1,0)$.

9.5. A taxicab driver moves between the airport A and two hotels B and C according to the following rules. If he is at the airport, he will be at one of the two hotels next with equal probability. If at a hotel then he returns to the airport with probability 3/4 and goes to the other hotel with probability 1/4. (a) Find the transition matrix for the chain. (b) Suppose the driver begins at the airport at time 0. Find the probability for each of his three possible locations at time 2 and the probability he is at hotel B at time 3.

9.6. The Markov chain associated with a manufacturing process may be described as follows: A part to be manufactured will begin the process by entering step 1. After step 1, 20% of the parts must be reworked, i.e., returned to step 1, 10% of the parts are thrown away, and 70% proceed to step 2. After step 2, 5% of the parts must be returned to the step 1, 10% to step 2, 5% are scrapped, and 80% emerge to be sold for a profit. (a) Formulate a four-state Markov chain with states 1, 2, 3, and 4 where 3 = a part that was scrapped and 4 = a part that was sold for a profit. (b) Compute the probability a part is scrapped in the production process.

9.7. Consider the following transition matrices. Which states are recurrent and which are transient? Give reasons for your answers.

(a)	1	2	3	4	5
1	.4	.3	.3	0	0
2	0	.5	0	.5	0
3	.5	0	.5	0	0
4	0	.5	0	.5	0
5	0	.3	0	.3	.4

(b)	1	2	3	4	5	6
1	.1	0	0	.4	.5	0
2	.1	.2	.2	0	.5	0
3	0	.1	.3	0	0	.6
4	.1	0	0	.9	0	0
5	0	0	0	.4	0	.6
6	0	0	0	0	.5	.5

9.8. Consider the following transition matrices. Identify the transient and recurrent states, and the irreducible closed sets in the Markov chains. Give reasons for your answers.

(a)	1	2	3	4	5
1	0	0	0	0	1
2	0	.2	0	.8	0
3	.1	.2	.3	.4	0
4	0	.6	0	.4	0
5	.3	0	0	0	.7

(b)	1	2	3	4	5	6
1	2/3	0	0	1/3	0	0
2	0	1/2	0	0	1/2	0
3	0	0	1/3	1/3	1/3	0
4	1/2	0	0	1/2	0	0
5	0	1/2	0	0	1/2	0
6	1/2	0	0	1/2	0	0

9.9. Six children (Dick, Helen, Joni, Mark, Sam, and Tony) play catch. If Dick has the ball he is equally likely to throw it to Helen, Mark, Sam, and Tony. If Helen has the ball she is equally likely to throw it to Dick, Joni, Sam, and Tony. If Sam has the ball he is equally likely to throw it to Dick, Helen, Mark, and Tony. If either Joni or Tony gets the ball, they keep throwing it to each other. If Mark gets the ball he runs away with it. (a) Find the transition probability and classify the states of the chain. (b) Suppose Dick has the ball at the beginning of the game. What is the probability Mark will end up with it?

9.10. *Brother–sister mating.* In this genetics scheme two individuals (one male and one female) are retained from each generation and are mated to give the next. If the individuals involved are diploid and we are interested in a trait with two alleles, A and a, then each individual has three possible states AA, Aa, aa or more succinctly 2, 1, 0. If we keep track of the sexes of the two individuals the chain has nine states, but if we ignore the sex there are just six: 22, 21, 20, 11, 10, and 00. (a) Assuming that reproduction corresponds to picking one letter at random from each parent, compute the transition probability. (b) 22 and 00 are absorbing states for the chain. Show that the probability of absorption in 22 is equal to the fraction of A's in the state. (c) Let $T = \min\{n \geq 0 : X_n = 22 \text{ or } 00\}$ be the absorption time. Find $E_x T$ for all states x.

9.11. Find the stationary distributions for the Markov chains with transition matrices:

(a)	1	2	3
1	.4	.6	0
2	.2	.4	.4
3	0	.3	.7

(b)	1	2	3	4
1	.4	.6	0	0
2	0	.7	.3	0
3	.1	0	.4	.5
4	.5	0	0	.5

9.12. Consider the Markov chain with transition matrix:

	1	2	3	4
1	0	0	3/5	2/5
2	0	0	1/5	4/5
3	1/4	3/4	0	0
4	1/2	1/2	0	0

(a) Compute p^2. (b) Find the stationary distributions of p and p^2. (c) Find the limit of $p^{2n}(x,x)$ as $n \to \infty$.

9.13. Three of every four trucks on the road are followed by a car, while only one of every five cars is followed by a truck. What fraction of vehicles on the road are trucks?

9.14. In a test paper the questions are arranged so that 3/4's of the time a True answer is followed by a True, while 2/3's of the time a False answer is followed by a False. You are confronted with a 100 question test paper. Approximately what fraction of the answers will be True?

9.15. In unprofitable times corporations sometimes suspend dividend payments. Suppose that after a dividend has been paid the next one will be paid with probability 0.9, while after a dividend is suspended the next one will be suspended with probability 0.6. In the long run what is the fraction of dividends that will be paid?

9.16. Folk wisdom holds that in Ithaca in the summer it rains 1/3 of the time, but a rainy day is followed by a second one with probability 1/2. Suppose that Ithaca weather is a Markov chain. What is its transition probability?

9.17. Consider a general chain with state space $S = \{1, 2\}$ and write the transition probability as

	1	2
1	$1-a$	a
2	b	$1-b$

Use the Markov property to show that

$$P(X_{n+1} = 1) - \frac{b}{a+b} = (1-a-b)\left\{P(X_n = 1) - \frac{b}{a+b}\right\}$$

and then conclude

$$P(X_n = 1) = \frac{b}{a+b} + (1-a-b)^n \left\{P(X_0 = 1) - \frac{b}{a+b}\right\}$$

This shows that if $0 < a+b < 2$, then $P(X_n = 1)$ converges exponentially fast to its limiting value $b/(a+b)$.

9.18. (a) Suppose brands A and B have consumer loyalties of .7 and .8, meaning that a customer who buys A one week will with probability .7 buy it again the next week, or try the other brand with .3. What is the limiting market share for each of these products? (b) Suppose now there is a third brand with loyalty .9, and that a consumer who changes brands picks one of the other two at random. What is the new limiting market share for these three products?

9.19. A certain town never has two sunny days in a row. Each day is classified as rainy, cloudy, or sunny. If it is sunny one day then it is equally likely to be cloudy or rainy the next. If it is cloudy or rainy, then it remains the same 1/2 of the time, but if it changes it will go to either of the other possibilities with probability 1/4 each. In the long run what proportion of days in this town are sunny? cloudy? rainy?

9.20. *Income classes.* In Example 1.8, we supposed that from one generation to the next families change their income group Low, Middle, or High according to the following Markov chain.

	L	M	H
L	0.6	0.3	0.1
M	0.2	0.7	0.1
H	0.1	0.3	0.6

Find the limiting fractions of the population in the three income classes.

9.21. A professor has two light bulbs in his garage. When both are burned out, they are replaced, and the next day starts with two working light bulbs. Suppose that when both are working, one of the two will go out with probability .02 (each has probability .01 and we ignore the possibility of losing two on the same day). However, when only one is there, it will burn out with probability .05. (i) What is the long-run fraction of time that there is exactly one bulb working? (ii) What is the expected time between light bulb replacements?

9.22. *Random walk on the circle.* Consider the points $1, 2, 3, 4$ to be marked on a ring. Let X_n be a Markov chain that moves to the right with probability p and to the left with probability $1-p$, subject to the rules that if X_n goes to the left from 1 it ends up at 4, and if X_n goes to the right from 4 it ends up at 1. Find (a) the transition probability for the chain, and (b) the limiting amount of time the chain spends at each site.

9.23. *Reflecting random walk on the line.* Consider the points $1, 2, 3, 4$ to be marked on a straight line. Let X_n be a Markov chain that moves to the right with probability 2/3 and to the left with probability 1/3, but subject this time to the rule that if X_n tries to go to the left from 1 or to the right from 4 it stays put. Find (a) the transition probability for the chain, and (b) the limiting amount of time the chain spends at each site.

9.24. A basketball player makes a shot with the following probabilities:

1/2 if he has missed the last two times
2/3 if he has hit one of his last two shots
3/4 if he has hit both of his last two shots

Formulate a Markov chain to model his shooting, and compute the limiting fraction of time he hits a shot.

9.25. An individual has three umbrellas, some at her office, and some at home. If she is leaving home in the morning (or leaving work at night) and it is raining, she will take an umbrella, if one is there. Otherwise, she gets wet. Assume that independent of the past, it rains on each trip with probability 0.2. To formulate a Markov chain, let X_n be the number of umbrellas at her current location. (a) Find the transition probability for this Markov chain. (b) Calculate the limiting fraction of time she gets wet.

9.26. Let X_n be the number of days since Mickey Markov last shaved, calculated at 7:30AM when he is trying to decide if he wants to shave today. Suppose that X_n is a Markov chain with transition matrix

	1	2	3	4
1	1/2	1/2	0	0
2	2/3	0	1/3	0
3	3/4	0	0	1/4
4	1	0	0	0

In words, if he last shaved k days ago, he will not shave with probability $1/(k+1)$. However, when he has not shaved for 4 days his wife orders him to shave, and he does so with probability 1. (a) What is the long-run fraction of time Mickey shaves? (b) Does the stationary distribution for this chain satisfy the detailed balance condition?

9.27. At the end of a month, a large retail store classifies each of its customer's accounts according to current (0), 30–60 days overdue (1), 60–90 days overdue (2), more than 90 days (3). Their experience indicates that the accounts move from state to state according to a Markov chain with transition probability matrix:

	0	1	2	3
0	.9	.1	0	0
1	.8	0	.2	0
2	.5	0	0	.5
3	.1	0	0	.9

In the long run what fraction of the accounts are in each category?

9.28. At the beginning of each day, a piece of equipment is inspected to determine its working condition, which is classified as state $1 =$ new, 2, 3, or $4 =$ broken. We assume the state is a Markov chain with the following transition matrix:

	1	2	3	4
1	.95	.05	0	0
2	0	.9	.1	0
3	0	0	.875	.125

(a) Suppose that a broken machine requires three days to fix. To incorporate this into the Markov chain we add states 5 and 6 and suppose that $p(4,5) = 1$, $p(5,6) = 1$, and $p(6,1) = 1$. Find the fraction of time that the machine is working. (b) Suppose now that we have the option of performing preventative maintenance when the machine is in state 3, and that this maintenance takes one day and returns the machine to state 1. This changes the transition probability to

	1	2	3
1	.95	.05	0
2	0	.9	.1
3	1	0	0

Find the fraction of time the machine is working under this new policy.

9.29. Let N_n be the number of heads observed in the first n flips of a fair coin and let $X_n = N_n$ mod 5, i.e., the remainder when we divide by 5. Use the Markov chain X_n to find the limit as $n \to \infty$ of $P(N_n$ is a multiple of 5).

9.30. *Landscape dynamics.* To make a crude model of a forest we might introduce states 0 = grass, 1 = bushes, 2 = small trees, 3 = large trees, and write down a transition matrix like the following:

	0	1	2	3
0	1/2	1/2	0	0
1	1/24	7/8	1/12	0
2	1/36	0	8/9	1/12
3	1/8	0	0	7/8

The idea behind this matrix is that if left undisturbed a grassy area will see bushes grow, then small trees, which of course grow into large trees. However, disturbances such as tree falls or fires can reset the system to state 0. Find the limiting fraction of land in each of the states.

9.31. *Wright–Fisher model.* Consider the chain described in Example 1.7.

$$p(x,y) = \binom{N}{y}(\rho_x)^y(1-\rho_x)^{N-y}$$

where $\rho_x = (1-u)x/N + v(N-x)/N$. (a) Show that if $u, v > 0$, then $\lim_{n\to\infty} p^n(x,y) = \pi(y)$, where π is the unique stationary distribution. There is no known formula for $\pi(y)$, but you can (b) compute the mean $\nu = \sum_y y\pi(y) = \lim_{n\to\infty} E_x X_n$.

9.32. *Ehrenfest chain.* Consider the Ehrenfest chain, Example 1.2, with transition probability $p(i, i+1) = (N-i)/N$, and $p(i, i-1) = i/N$ for $0 \le i \le N$.

Let $\mu_n = E_x X_n$. (a) Show that $\mu_{n+1} = 1 + (1 - 2/N)\mu_n$. (b) Use this and induction to conclude that

$$\mu_n = \frac{N}{2} + \left(1 - \frac{2}{N}\right)^n (x - N/2)$$

From this we see that the mean μ_n converges exponentially rapidly to the equilibrium value of $N/2$ with the error at time n being $(1 - 2/N)^n(x - N/2)$.

9.33. Roll a fair die repeatedly and let $Y_1, Y_2, \ldots$ be the resulting numbers. Let $X_n = |\{Y_1, Y_2, \ldots, Y_n\}|$ be the number of values we have seen in the first n rolls for $n \geq 1$ and set $X_0 = 0$. X_n is a Markov chain. (a) Find its transition probability. (b) Let $T = \min\{n : X_n = 6\}$ be the number of trials we need to see all 6 numbers at least once. Find ET.

9.34. *Coupon collector's problem.* We are interested now in the time it takes to collect a set of N baseball cards. Let T_k be the number of cards we have to buy before we have k that are distinct. Clearly, $T_1 = 1$. A little more thought reveals that if each time we get a card chosen at random from all N possibilities, then for $k \geq 1$, $T_{k+1} - T_k$ has a geometric distribution with success probability $(N-k)/N$. Use this to show that the mean time to collect a set of N baseball cards is $\approx N \log N$, while the variance is $\approx N^2 \sum_{k=1}^{\infty} 1/k^2$.

9.35. If we have a deck of 52 cards, then its state can be described by a sequence of numbers that gives the cards we find as we examine the deck from the top down. Since all the cards are distinct, this list is a permutation of the set $\{1, 2, \ldots 52\}$, i.e., a sequence in which each number is listed exactly once. Formulas from elementary probability tells us there are $52! = 52 \cdot 51 \cdots 2 \cdot 1$ possible permutations. Consider the following very simple *shuffling procedure* that is easy to implement on a computer: pick a card at random from the deck (including possibly the top card) and put it on top. Show that if we repeatedly apply this algorithm, then in the limit the deck is perfectly shuffled in the sense that all 52! possibilities are equally likely.

9.36. *Bernoulli–Laplace model of diffusion.* Consider two urns each of which contains m balls; b of these $2m$ balls are black, and the remaining $2m - b$ are white. We say that the system is in state i if the first urn contains i black balls and $m - i$ white balls while the second contains $b - i$ black balls and $m - b + i$ white balls. Each trial consists of choosing a ball at random from each urn and exchanging the two. Let X_n be the state of the system after n exchanges have been made. X_n is a Markov chain. (a) Compute its transition probability. (b) Verify that the stationary distribution is given by

$$\pi(i) = \binom{b}{i}\binom{2m-b}{m-i} \bigg/ \binom{2m}{m}$$

(c) Can you give a simple intuitive explanation why the formula in (b) gives the right answer?

9.37. Library chain. On each request the ith of n possible books is the one chosen with probability p_i. To make it quicker to find the book the next time, the librarian moves the book to the left end of the shelf. Define the state at any time to be the sequence of books we see as we examine the shelf from left to right. Since all the books are distinct this list is a permutation of the set $\{1, 2, \ldots n\}$, i.e., each number is listed exactly once. Show that

$$\pi(i_1, \ldots, i_n) = p_{i_1} \cdot \frac{p_{i_2}}{1 - p_{i_1}} \cdot \frac{p_{i_3}}{1 - p_{i_1} - p_{i_2}} \cdots \frac{p_{i_n}}{1 - p_{i_1} - \cdots p_{i_{n-1}}}$$

is a stationary distribution.

9.38. A criminal named Xavier and a policeman named Yakov move between three possible hideouts according to Markov chains X_n and Y_n with transition probabilities:

$$p_{Xavier} = \begin{pmatrix} .6 & .2 & .2 \\ .2 & .6 & .2 \\ .2 & .2 & .6 \end{pmatrix} \quad \text{and} \quad p_{Yakov} = \begin{pmatrix} 0 & .5 & .5 \\ .5 & 0 & .5 \\ .5 & .5 & 0 \end{pmatrix}$$

At time $T = \min\{n : X_n = Y_n\}$ the game is over and the criminal is caught.
(a) Suppose $X_0 = i$ and $Y_0 = j \neq i$. Find the expected value of T. (b) Suppose that the two players generalize their strategies to

$$p_{Xavier} = \begin{pmatrix} 1 - 2p & p & p \\ p & 1 - 2p & p \\ p & p & 1 - 2p \end{pmatrix}$$

$$p_{Yakov} = \begin{pmatrix} 1 - 2q & q & q \\ q & 1 - 2q & q \\ q & q & 1 - 2q \end{pmatrix}$$

If Yakov uses $q = 0.5$ as he did in part (a) what value of p should Xavier choose to maximize the expected time to get caught? Answer the last question again for $q = 1/3$.

9.39. In tennis the winner of a game is the first player to win four points, unless the score is $4 - 3$, in which case the game must continue until one player wins by two points. (a) Suppose the game has reached the point where the score is tied at three points (or more) each and that the server will independently win the point with probability 0.6. What is the probability the server will win the game? (b) What is the probability the server will win if she is ahead by one point? (c) Behind by one point?

9.40. At the New York State Fair in Syracuse, David Durrett encounters a carnival game where for one dollar he may buy a single coupon allowing him to play the game of *Dummo*. On each play of *Dummo*, David has an even chance of winning or losing a coupon. When he runs out of coupons he loses the game. However, if he can collect three coupons, he wins a surprise. (a) What is the probability David will win the suprise? (b) What is the expected number of plays he needs to win or lose the game. (c) Answer (a) and (b) when the goal is N coupons.

9.41. A warehouse has a capacity to hold four items. If the warehouse is neither full nor empty, the number of items in the warehouse changes whenever a new item is produced or an item is sold. Suppose that (no matter when we look) the probability that the next event is "a new item is produced" is $2/3$ and that the new event is a "sale" is $1/3$. If there is currently one item in the warehouse, what is the probability that the warehouse will become full before it becomes empty.

9.42. Consider roulette as treated in Example 6.3, but this time imagine we are the house, winning each bet with probability $20/38$ and losing with probability $18/38$. Suppose now that we start our casino with \$200 capital and invite people one at a time to bet \$1 on our roulette wheel. What is the probability we will ever go bankrupt?

9.43. *Random walk on a clock.* Consider the numbers $1, 2, \ldots 12$ written around a ring as they usually are on a clock. Consider a Markov chain that at any point jumps with equal probability to the two adjacent numbers. (a) What is the expected number of steps that X_n will take to return to its starting position? (b) What is the probability X_n will visit all the other states before returning to its starting position?

9.44. *Knight's random walk.* If we represent our chessboard as $\{(i,j) : 1 \leq i, j \leq 8\}$ then a knight can move from (i,j) to any of eight squares $(i+2, i+1)$, $(i+2, i-1)$, $(i+1, i+2)$ $(i+1, i-2)$, $(i-1, i+2)$, $(i-1, i-2)$, $(i-2, i+1)$, or $(i-2, i-1)$, provided of course that they are on the chessboard. Let X_n be the sequence of squares that results if we pick one of knights legal moves at random. Find (a) the stationary distribution and (b) the expected number of moves to return to corner (1,1) when we start there.

9.45. *Queen's random walk.* Again we represent our chessboard as $\{(i,j) : 1 \leq i, j \leq 8\}$. A queen can move any number of squares horizontally, vertically, or diagonally. Let X_n be the sequence of squares that results if we pick one of queen's legal moves at random. Find (a) the stationary distribution and (b) the expected number of moves to return to corner (1,1) when we start there.

9.46. *General birth and death chains.* The state space is $\{0, 1, 2, \ldots\}$ and the transition probability has

$$p(x, x+1) = p_x$$
$$p(x, x-1) = q_x \quad \text{for } x > 0$$
$$p(x, x) = r_x \quad \text{for } x \geq 0$$

while the other $p(x,y) = 0$. Let $V_y = \min\{n \geq 0 : X_n = y\}$ be the time of the first visit to y and let $h_N(x) = P_x(V_N < V_0)$. By considering what happens on the first step, we can write

$$h_N(x) = p_x h_N(x+1) + r_x h_N(x) + q_x h_N(x-1)$$

Set $h_N(1) = c_N$ and solve this equation to conclude that 0 is recurrent if and only if $\sum_{y=1}^{\infty} \prod_{x=1}^{y-1} q_x/p_x = \infty$ where by convention $\prod_{x=1}^{0} = 1$.

9.47. To see what the conditions in the last problem say we will now consider some concrete examples. Let $p_x = 1/2$, $q_x = e^{-cx^{-\alpha}}/2$, $r_x = 1/2 - q_x$ for $x \geq 1$ and $p_0 = 1$. For large x, $q_x \approx (1 - cx^{-\alpha})/2$, but the exponential formulation keeps the probabilities nonnegative and makes the problem easier to solve. Show that the chain is recurrent if $\alpha > 1$ or if $\alpha = 1$ and $c \leq 1$ but is transient otherwise.

9.48. Consider the Markov chain with state space $\{0, 1, 2, \ldots\}$ and transition probability

$$p(m, m+1) = \frac{1}{2}\left(1 - \frac{1}{m+2}\right) \quad \text{for } m \geq 0$$
$$p(m, m-1) = \frac{1}{2}\left(1 + \frac{1}{m+2}\right) \quad \text{for } m \geq 1$$

and $p(0,0) = 1 - p(0,1) = 3/4$. Find the stationary distribution π.

9.49. Consider the Markov chain with state space $\{1, 2, \ldots\}$ and transition probability

$$p(m, m+1) = m/(2m+2) \quad \text{for } m \geq 1$$
$$p(m, m-1) = 1/2 \quad \text{for } m \geq 2$$
$$p(m, m) = 1/(2m+2) \quad \text{for } m \geq 2$$

and $p(1,1) = 1 - p(1,2) = 3/4$. Show that there is no stationary distribution.

9.50. Consider the aging chain on $\{0, 1, 2, \ldots\}$ in which for any $n \geq 0$ the individual gets one day older from n to $n+1$ with probability p_n but dies and returns to age 0 with probability $1 - p_n$. Find conditions that guarantee that (a) 0 is recurrent, (b) positive recurrent. (c) Find the stationary distribution.

9.51. The opposite of the aging chain is the renewal chain with state space $\{0, 1, 2, \ldots\}$ in which $p(i, i-1) = 1$ when $i > 0$. The only nontrivial part of the transition probability is $p(0, i) = p_i$. Show that this chain is always recurrent but is positive recurrent if and only if $\sum_n n p_n < \infty$.

9.52. Consider a branching process as defined in Example 7.2, in which each family has exactly three children, but invert Galton and Watson's original motivation and ignore male children. In this model a mother will have an average of 1.5 daughters. Compute the probability that a given woman's descendents will die out.

9.53. Consider a branching process as defined in Example 7.2, in which each family has a number of children that follows a shifted geometric distribution: $p_k = p(1-p)^k$ for $k \geq 0$, which counts the number of failures before the first success when success has probability p. Compute the probability that starting from one individual the chain will be absorbed at 0.

2 Martingales

In this chapter we will introduce a class of process that can be thought of as the fortune of a gambler betting on a fair game. We will not go very far with their theory, proving only the "optional stopping theorem." This theorem is a useful tool for doing computations. In particular, it will help us to see easily some of the results we found in Section 1.6, and it will be very useful when we come to Brownian motion in Chapter 6.

2.1. Conditional Expectation

Our study of martingales will rely heavily on the notion of conditional expectation and involve some formulas that may not be familiar, so we will review them here. We begin with several definitions. Given an event A we define its **indicator function**

$$1_A = \begin{cases} 1 & x \in A \\ 0 & x \in A^c \end{cases}$$

In words, 1_A is "1 on A" (and 0 otherwise). Given a random variable Y, we define the **integral of Y over A** to be

$$E(Y;A) = E(Y1_A)$$

Note that on the right multiplying by 1_A sets the product $= 0$ on A^c and leaves the values on A unchanged. Finally, we define the **conditional expectation of Y given A** to be

$$E(Y|A) = E(Y;A)/P(A)$$

This is, of course, the expected value for the probability defined by

$$P(\cdot|A) = P(\cdot \cap A)/P(A)$$

It is easy to see from the definition that the integral over A is linear:

(1.1) $$E(Y+Z;A) = E(Y;A) + E(Z;A)$$

so dividing by $P(A)$, conditional expectation also has this property

(1.2) $$E(Y+Z|A) = E(Y|A) + E(Z|A)$$

(Provided of course that all of the expected values exist.) In addition, as in ordinary integration one can take constants outside of the integral.

(1.3) Lemma. *If X is a constant c on A, then $E(XY|A) = cE(Y|A)$.*

Proof. Since $X = c$ on A, $XY1_A = cY1_A$. Taking expected values and pulling the constant out front, $E(XY1_A) = E(cY1_A) = cE(Y1_A)$. Dividing by $P(A)$ now gives the result. □

Our last two properties concern the behavior of $E(Y;A)$ and $(Y|A)$ as a function of the set A.

(1.4) Lemma. *If B is the disjoint union of $A_1, \ldots, A_k$, then*

$$E(Y;B) = \sum_{j=1}^{k} E(Y;A_j)$$

Proof. Our assumption implies $Y1_B = \sum_{j=1}^{k} Y1_{A_j}$, so taking expected values, we have

$$E(Y;B) = E(Y1_B) = E\left(\sum_{j=1}^{k} Y1_{A_j}\right) = \sum_{j=1}^{k} E(Y1_{A_j}) = \sum_{j=1}^{k} E(Y;A_j) \quad \square$$

(1.5) Lemma. *If B is the disjoint union of $A_1, \ldots, A_k$, then*

$$E(Y|B) = \sum_{j=1}^{k} E(Y|A_j) \cdot \frac{P(A_j)}{P(B)}$$

Proof. Using the definition of conditional expectation, formula (1.4), then doing some arithmetic and using the definition again, we have

$$E(Y|B) = E(Y;B)/P(B) = \sum_{j=1}^{k} E(Y;A_j)/P(B)$$

$$= \sum_{j=1}^{k} \frac{E(Y;A_j)}{P(A_j)} \cdot \frac{P(A_j)}{P(B)} = \sum_{j=1}^{k} E(Y|A_j) \cdot \frac{P(A_j)}{P(B)} \quad \square$$

2.2. Examples of Martingales

We begin by giving the definition of a martingale. Thinking of M_n as the amount of money at time n for a gambler betting on a fair game, we say that $M_0, M_1, \ldots$ is a **martingale** if for any $n \geq 0$ we have $E|M_n| < \infty$ and for any possible values $m_n, \ldots, m_0$

(2.1) $\qquad E(M_{n+1}|M_n = m_n, M_{n-1} = m_{n-1}, \ldots M_0 = m_0) = m_n$

The first condition, $E|M_n| < \infty$, is needed to guarantee that the conditional expectation makes sense. The second, defining property says that conditional on the past up to time n the average amount of money the gambler has at time $n+1$ is what he has at time n. To relate this to fair games we can rewrite (2.1) as

(2.2) $\qquad E(M_{n+1} - M_n|M_n = m_n, M_{n-1} = m_{n-1}, \ldots M_0 = m_0) = 0$

In words, the expected winnings on the next play, $M_{n+1} - M_n$, have mean zero conditional on the past.

To explain the reason for our interest in martingales, we will now give a number of examples.

Example 2.1. Mean zero random walks. Let $\xi_1, \xi_2, \ldots$ be independent with $E\xi_i = 0$. Note that we have not assumed that all the ξ_i have the same distribution. $M_n = M_0 + \xi_1 + \cdots + \xi_n$ defines a martingale.

Proof. To check this, note that $M_{n+1} - M_n = \xi_{n+1}$, which is independent of $M_0, \ldots, M_n$, so the conditional mean of the difference is just the mean:

$$E(M_{n+1} - M_n|M_n = m_n, M_{n-1} = m_{n-1}, \ldots, M_0 = m_0) = E\xi_{n+1} = 0 \qquad \square$$

In most cases, casino games are not fair but biased against the player. We say that Y_n is a **supermartingale** if a gambler's expected winnings on one play are negative:

$$E(Y_{n+1} - Y_n|Y_n = y_n, Y_{n-1} = y_{n-1}, \ldots, Y_0 = y_0) \leq 0$$

If we reverse the sign and suppose

$$E(Y_{n+1} - Y_n|Y_n = y_n, Y_{n-1} = y_{n-1}, \ldots, Y_0 = y_0) \geq 0$$

then Y_n is called a **submartingale**. An obvious modification of Example 1.1 shows that if $\xi_1, \xi_2, \ldots$ are independent with $E\xi_i \leq 0$, then $Y_n = Y_0 + \xi_1 + \cdots + \xi_n$

Section 2.2 Examples of Martingales

defines a supermartingale, while if all the $E\xi_i \geq 0$, then Y_n is a submartingale. Of course, given any independent $\xi_1, \xi_2, \ldots$ with finite means we can define a martingale by

$$M_n = M_0 + (\xi_1 - E\xi_1) + \cdots + (\xi_n - E\xi_n)$$

For another example of the three types of processes consider

Example 2.2. Branching processes. Consider a population in which each individual in the nth generation gives birth to an independent and identically number of children. Let Z_n be the number of individuals at time n. Suppose that k children are produced with probability p_k and let $\mu = \sum_k k p_k$ be the mean number of children. Since each of the Z_n members of the nth generation will have an average of μ children, it is clear that

$$E(Z_{n+1}|Z_n = z_n, Z_{n-1} = z_{n-1}, \ldots, Z_0 = z_0) = \mu z_n$$

so Z_n is a supermartingale if $\mu \leq 1$, a martingale if $\mu = 1$, and a submartingale if $\mu \geq 1$. Dividing each side of the last equation by μ^n gives

$$E\left(\frac{Z_{n+1}}{\mu^{n+1}} \bigg| Z_n = z_n, Z_{n-1} = z_{n-1}, \ldots, Z_0 = z_0\right) = \frac{z_n}{\mu^n}$$

so $M_n = Z_n/\mu^n$ is a martingale. □

If you would rather watch your money grow instead, consider:

Example 2.3. Stock prices. Let $\zeta_1, \zeta_2, \ldots$ be independent positive random variables with $E\zeta_i < \infty$ and let $X_n = X_0 \cdot \zeta_1 \cdots \zeta_n$. Here, we think of $\zeta_n - 1$ as the change in the value of the stock over a fixed time interval (e.g., one day or one week) as a fraction of its current value. This multiplicative formulation is natural since it easily guarantees that stock prices are nonnegative and empirical evidence suggests that fluctuations in the value of a stock are roughly proportional to its price. Two concrete examples are:

Discrete Black–Scholes model. $\zeta_i = e^{\eta_i}$ where $\eta_i = \text{normal}(\mu, \sigma^2)$, that is, normal with mean μ and variance σ^2.

Binomial model. $\zeta_i = (1+a)e^{-r}$ with probability p and $(1+a)^{-1}e^{-r}$. Here r is interest rate by which we discount future rewards, while the use of $(1+a)$ and $(1+a)^{-1}$ guarantees that at time n the price has the form $X_0(1+a)^k e^{-nr}$.

Writing $X_{n+1} = X_n \zeta_{n+1}$ and using (1.3), we have

$$E(X_{n+1}|X_n = x_n, \ldots, X_0 = x_0)$$
$$= x_n E(\zeta_{n+1}|X_n = x_n, \ldots, X_0 = x_0) = x_n E\zeta_{n+1}$$

Since ζ_{n+1} is independent of $X_0, \ldots, X_n$. Thus X_n is a submartingale if $E\zeta_i > 1$, a martingale if $E\zeta_i = 1$, and a supermartingale if $E\zeta_i < 1$. □

Having seen three examples of our three types of processes, we will pause to introduce three theoretical results. Thinking of a supermartingale as the fortune of a gambler playing an unfavorable game, it is natural to guess that although he may occasionally win, the expected value of his fortune is decreasing in time.

(2.3) Theorem. *If Y_m is a supermartingale then $EY_m \geq EY_n$.*

Proof. It is enough to show that the expected value decreases with each time step, i.e., $EY_k \geq EY_{k+1}$. To do this, we write i as shorthand for the vector $(i_n, i_{n-1}, \ldots i_0)$, let

$$A_i = \{Y_n = i_n, Y_{n-1} = i_{n-1}, \ldots, Y_0 = i_0\}$$

and note that (1.4) and the definition of conditional expectation imply

$$E(Y_{k+1} - Y_k) = \sum_i E(Y_{k+1} - Y_k; A_i)$$
$$= \sum_i P(A_i) E(Y_{k+1} - Y_k | A_i) \leq 0$$

since each term $E(Y_{k+1} - Y_k | A_i) \leq 0$. □

The result in (2.3) generalizes immediately to our other two types of processes. Multiplying by -1 we see:

(2.4) Theorem. *If Y_m is a submartingale and $0 \leq m < n$, then $EY_m \leq EY_n$.*

Since a process is a martingale if and only if it is both a supermartingale and submartingale, we can conclude that:

(2.5) Theorem. *If M_m is a martingale and $0 \leq m < n$ then $EM_m = EM_n$.*

In Section 1.6 we have seen several examples of martingales.

Example 2.4. Wright–Fisher model with no mutation. As we discussed in Example 6.4, this chain X_n has state space $S = \{0, 1, \ldots N\}$ and transition probability

$$p(x, y) = \binom{N}{y} \left(\frac{x}{N}\right)^y \left(\frac{N-x}{N}\right)^{N-y}$$

The right hand side is the binomial($N, x/N$) distribution, i.e., the number of successes in N trials when success has probability x/N, so the mean number of successes is x. That is, we have

$$E(X_{n+1}|X_n = x_n, X_{n-1} = x_{n-1}, \ldots, X_0 = x_0) = x_n$$

i.e., X_n is martingale. You will see in Exercise 4.1 (on page 115) that this is the reason behind the result in (6.1) of Chapter 1

$$P_x(V_N < V_0) = x/N$$

Intuitively, (2.5) implies that the average number of A's is constant in time, so if we start out with x of the A genes, then the probability we end up with N of them instead of 0 must be x/N to make the mean x at the end.

Abstracting from the last example leads us to a general result.

(2.6) Theorem. *Let X_n be a Markov chain with transition probability p and let $h(x, n)$ be a function of the state x and the time n so that*

$$h(x, n) = \sum_y p(x, y) h(y, n+1)$$

Then $M_n = h(X_n, n)$ is a martingale.

In Example 2.4, the function $h(x, n) = x$ does not depend on time. However we will soon want to apply this to $h(x, n) = x^2 - cn$ and $h(x, n) = \exp(\theta x - cn)$ for suitably chosen values of c.

Proof. By the Markov property and our assumption on h

$$E(h(X_{n+1}, n+1)|X_n = x_n, \ldots, X_0 = x_0) = \sum_y p(x_n, y) h(y, n+1) = h(x_n, n)$$

What we have shown is not quite the definition of martingale, which is

$$E(M_{n+1}|M_n = m_n, \ldots, M_0 = m_0) = m_n$$

To bridge the gap fix $m_n, \ldots, m_0$ and let $B = \{M_n = m_n, \ldots, M_0 = m_0\}$. Since $M_n = f(X_n, n)$ the set B can be written as a disjoint union $B = \cup_{i \in I} A_i$ where $A_i = \{X_n = i_n, \ldots, X_0 = i_0\}$ and we have written i as shorthand for the vector $(i_0, \ldots, i_n)$. Using (1.5) now and $h(x_n, n) = m_n$), we have

$$E(M_{n+1}|B) = \sum_{i \in I} E(M_{n+1}|A_i) \cdot \frac{P(A_i)}{P(B)} = m_n \sum_{i \in I} \frac{P(A_i)}{P(B)} = m_n$$

106 Chapter 2 Martingales

since each $E(M_{n+1}|A_i) = m_n$ and $\sum_{i \in I} P(A_i) = P(B)$. □

The last proof motivates the following "improved" definition. We have used quotes since the new definition leads to more useful theoretical results but is more mysterious.

We say that $M_0, M_1, \ldots$ is a **martingale with respect to** $X_0, X_1, \ldots$ if for any $n \geq 0$ we have $E|M_n| < \infty$ and for any possible values $x_n, \ldots, x_0$

(2.7) $\quad E(M_{n+1} - M_n | X_n = x_n, X_{n-1} = x_{n-1}, \ldots X_0 = x_0) = 0$

In all of our examples X_n will be a Markov chain and $M_n = h(X_n, n)$. The proof of (2.6) shows that if M_n is a martingale with respect to some sequence X_n then it is an ordinary martingale, i.e., as defined in (2.1). The real reason for our interest in replacing (2.1) by (2.7) won't become clear until we apply the optional stopping theorem to examples in Section 2.4. There the new definition will have the advantage that our "stopping times" can be based on the values of $X_n, \ldots, X_0$ rather than just on $M_n, \ldots, M_0$. To see that this can make a difference consider:

Example 2.5. Variance martingale. Let $\xi_1, \xi_2, \ldots$ be independent with $E\xi_i = 0$ and variance $E\xi_i^2 = \sigma_i^2$. Again we have not assumed that all the ξ_i have the same distribution. Let $S_n = S_0 + \xi_1 + \cdots + \xi_n$, where S_0 is a constant, and let $v_n = \sum_{i=1}^{n} \sigma_i^2$ be the variance of S_n.

(2.8) $\quad M_n = S_n^2 - v_n$ is a martingale with respect to S_n.

Proof. Taking differences, $v_{n+1} - v_n = \sigma_{n+1}^2$, so

$$M_{n+1} - M_n = (S_n + \xi_{n+1})^2 - S_n^2 - \sigma_{n+1}^2 = 2\xi_{n+1}S_n + \xi_{n+1}^2 - \sigma_{n+1}^2$$

If we let $A = \{S_n = s_n, \ldots, S_0 = s_0\}$, then since $S_n = s_n$ on A, (1.3) implies

$$E(2\xi_{n+1}S_n | A) = 2s_n E(\xi_{n+1}|A) = 0$$

Since ξ_{n+1} is independent of A and has mean 0 and variance σ_{n+1}^2, we have

$$E(\xi_{n+1}^2 - \sigma_{n+1}^2 | A) = 0$$

Combining the last three equalities shows that $E(M_{n+1} - M_n | A) = 0$. □

In Section 2.4 we will use this result to derive (6.4) from Chapter 1, which gives the expected duration of fair games: if S_n is a symmetric simple random walk and $\tau = \min\{n : S_n \notin (0, N)\}$, then $E_x \tau = x(N - x)$. The time τ can be determined by watching the sequence S_n but not from watching $M_n = S_n^2 - n$, since squaring S_n leaves us unable to distinguish $S_n = N$ from $S_n = -N$.

Our next example is also from Section 1.6.

Example 2.6. Gambler's ruin. Let $\xi_1, \xi_2, \ldots$ be independent with
$$P(\xi_i = 1) = p \quad \text{and} \quad P(\xi_i = -1) = q = 1 - p$$
Let $S_n = S_0 + \xi_1 + \cdots + \xi_n$ and $g(x) = ((1-p)/p)^x$.

(2.9) $M_n = g(S_n)$ is a martingale with respect to S_n.

Proof. Using (2.6) with $h(x, n) = g(x)$, we need only check that $g(x) = \sum_y p(x,y)g(y)$. To do this we note that

$$\sum_y p(x,y)g(y) = p \cdot \left(\frac{1-p}{p}\right)^{x+1} + (1-p) \cdot \left(\frac{1-p}{p}\right)^{x-1}$$

$$= (1-p) \cdot \left(\frac{1-p}{p}\right)^x + p \cdot \left(\frac{1-p}{p}\right)^x = \left(\frac{1-p}{p}\right)^x \qquad \square$$

The last example generalizes easily to give:

Example 2.7. Exponential martingale. Let $\xi_1, \xi_2, \ldots$ be independent and identically distributed with $\varphi(\theta) = E \exp(\theta \xi_1) < \infty$. Let $S_n = S_0 + \xi_1 + \cdots + \xi_n$. Then

(2.10) $M_n = \exp(\theta S_n)/\varphi(\theta)^n$ is a martingale with respect to S_n.

Proof. Since $S_{n+1} = S_n + \xi_{n+1}$ and S_n is constant on $\{S_n = s_n, \ldots S_0 = s_0\}$, (1.3) implies that

$$E(M_{n+1}|S_n = s_n, \ldots S_0 = s_0)$$
$$= \frac{\exp(\theta S_n)}{\varphi(\theta)^{n+1}} E(\exp(\xi_{n+1})|S_n = s_n, \ldots S_0 = s_0)$$
$$= \frac{\exp(\theta S_n)}{\varphi(\theta)^{n+1}} \cdot \varphi(\theta) = M_n \qquad \square$$

To see that Example 2.6 is a special case, note that when $P(\xi_i = 1) = p$ and $P(\xi_i = -1) = 1 - p$,
$$\varphi(\theta) = Ee^{\theta \xi_i} = p \cdot e^\theta + (1-p) \cdot e^{-\theta}$$

Chapter 2 Martingales

If we pick $e^\theta = (1-p)/p$, then $\varphi(\theta) = (1-p) + p = 1$ and the exponential martingale simplifies to

$$M_n = \exp(\theta S_n) = ((1-p)/p)^{S_n}$$

Closely related to the last example are a collection of martingales that are important in the theory of testing statistical hypotheses.

Example 2.8. Likelihood ratios. Suppose that we are testing the hypothesis that observations $\xi_1, \xi_2, \ldots$ are independent and have density function f but the truth is that $\xi_1, \xi_2, \ldots$ are independent and have density function g where $\{x : f(x) > 0\} = \{x : g(x) > 0\}$. Let

$$h(x) = \begin{cases} f(x)/g(x) & \text{when } g(x) > 0 \\ 0 & \text{when } g(x) = 0 \end{cases}$$

(2.11) $M_n = h(\xi_1) \cdots h(\xi_n)$ is a martingale with respect to ξ_n.

Note that the condition $\{f > 0\} \subset \{g > 0\}$ is needed for our statistical problem to make sense. Otherwise one observation landing in $\{f > 0, g = 0\}$ or $\{f = 0, g > 0\}$ will allow us to distinguish the two possibilities with certainty.

Proof. This is a special case of Example 2.3. Take $M_0 = 1$ and $\zeta_i = h(\xi_i)$, then note that

$$E\zeta_i = \int \frac{f(x)}{g(x)} \cdot g(x)\, dx = \int f(x)\, dx = 1$$

since f is a density function.   □

In our last three examples the martingales were products of random variables. Continuing that trend we have:

Example 2.9. Option prices. As in Example 2.3, let $\zeta_1, \zeta_2, \ldots$ be independent positive random variables that all have the same distribution, and define a stock price process by

$$X_n = X_0 \cdot \zeta_1 \cdots \zeta_n$$

Fix a terminal time N and introduce a payoff function $f(x)$. A common one is the "European call option":

$$f(x) = \begin{cases} (x-K) & \text{if } x > K \\ 0 & \text{if } x \leq K \end{cases}$$

Section 2.2 Examples of Martingales

In words, you have a right to buy the stock at a price K at time N. If the stock price $x > K$ you buy it and make a profit of $x - K$. If the stock price is $x < K$ you don't buy and your profit is 0. To write this more compactly it is useful to introduce the **positive part** of a real number y:

$$y^+ = \max\{y, 0\} = \begin{cases} y & \text{if } y > 0 \\ 0 & \text{if } y \leq 0 \end{cases}$$

and write $f(x) = (x - K)^+$.

Let $h(x, n) = E(f(X_N)|X_n = x)$ be the expected payoff at time N when the stock price is x at time $n < N$. By considering what happens on the first step, it is easy to see that

$$h(x, n) = Eh(X_{n+1}, n+1)$$

so (2.6) implies that $h(X_n, n)$ is a martingale. □

Our next martingale takes us in a different direction.

Example 2.10. Polya's urn scheme. Consider an urn that contains red and green balls. At time 0 there is one ball of each color. At time n we draw out a ball chosen at random. We return it to the urn and add one more ball of the color chosen. Let X_n be the fraction of red balls at time n. Since we add exactly one ball each time, there are always $n + 2$ balls at time n and hence if $X_n = x_n$ there are $(n + 2)x_n$ red balls. From this it follows that

$$E(X_{n+1}|X_n = x_n, \ldots X_0 = x_0) = x_n \cdot \frac{(n+2)x_n + 1}{n+3} + (1 - x_n) \cdot \frac{(n+2)x_n}{n+3}$$

$$= \frac{x_n}{n+3} + \frac{(n+2)x_n}{n+3} = x_n$$

so X_n is a martingale (with respect to X_n). □

In our final example we turn time around.

Example 2.11. Backwards random walk. Let $\xi_1, \xi_2, \ldots$ be independent and identically distributed with finite mean and let $S_n = \xi_1 + \cdots + \xi_n$. In this example we will look at $M_m = S_{n-m}/(n-m)$, $0 \leq m < n$ or more visually:

$$\frac{S_n}{n}, \frac{S_{n-1}}{n-1}, \ldots, \frac{S_2}{2}, \frac{S_1}{1}$$

(2.12) If we fix n, then $M_m = S_{n-m}/(n-m)$, $0 \leq m < n$ is a martingale (with respect to M_m).

Chapter 2 Martingales

Proof. It is clear from symmetry that if $i < j \le k$, then $E(\xi_i|S_k) = E(\xi_j|S_k)$. Summing the last equality from $j = 1$ to k and using (1.2) gives

$$kE(\xi_i|S_k) = \sum_{j=1}^{k} E(\xi_j|S_k) = E(S_k|S_k) = S_k$$

or $E(\xi_i|S_k) = S_k/k$, which is reasonable since ξ_i is one of k terms that makes up S_k. From the last equality we see that

$$E(S_{k-1}|S_k) = E(S_k - \xi_k|S_k) = S_k\left(1 - \frac{1}{k}\right) = \frac{S_k}{k}\cdot(k-1)$$

or $E(S_{k-1}/(k-1)|S_k) = S_k/k$. The variables $S_{k+1} - S_k, S_{k+2} - S_k, \cdots S_n - S_k$ are independent of $\xi_1, \ldots \xi_k$ so they contain no information about $S_{k-1}/(k-1)$ and we have

$$E(S_{k-1}/(k-1)|S_k, \ldots, S_n) = S_k/k$$

Letting $k = n - m$ and changing variables, we have

$$E(M_{m+1}|M_m, \ldots, M_0) = M_m$$

2.3. Optional Stopping Theorem

The most famous result of martingale theory is that "you can't make money playing a fair game" and hence "you can't beat an unfavorable game." In this section we will prove two results that make these statements precise. Our first step is analyze a famous gambling system and show why it doesn't work.

Example 3.1. Doubling strategy. Suppose you are playing a game in which you will win or lose $1 on each play. If you win you bet $1 on the next play but if you lose then you bet twice the previous amount. The idea behind the system can be seen by looking at what happens if we lose four times in a row then win:

outcome	L	L	L	L	W
bet	1	2	4	8	16
net profit	-1	-3	-7	-15	1

In this example our net profit when we win is $1. Since $1+2+\cdots+2^k = 2^{k+1}-1$, this is true if we lose k times in a row. Thus every time we win our net profit is up by $1 from the previous time we won.

Section 2.3 Optional Stopping Theorem

This system will succeed in making us rich as long as the probability of winning is positive, so where's the catch? The problem is this:

(3.1) Theorem. *Suppose we use the doubling system on a supermartingale up to a fixed time n and we let W_n be our net winnings. Then $EW_n \leq 0$.*

To prove this we will introduce a family of betting strategies that generalize the doubling strategy. The amount of money we bet on the nth game, H_n, clearly, cannot depend on the outcome of that game nor is it sensible to allow it to depend on the outcomes of games that will be played later. However, H_n can certainly be allowed to be any function of the previous outcomes $X_{n-1}, X_{n-2}, \ldots, X_0$. To say this formally,

Definition. Let $\mathcal{F}(X_0, \ldots, X_m)$ be the collection of random variables that can be written as a function $g(X_0, \ldots, X_m)$.

Assumption. We require that $H_n \in \mathcal{F}(X_0, \ldots, X_{n-1})$.

The last condition should be read as "H_n is a function of $X_0, X_1, \ldots X_{n-1}$."

To explain how to compute the payoff obtained from using a betting strategy, we begin with a simple example. Suppose the game is flipping a (possibly biased) coin and H_m is the amount we bet on heads at time m. Letting $\xi_m = 1$ if the mth coin flip is heads and -1 if the mth flip is tails, then $H_m \xi_m$ is our profit (or loss if negative) on the mth play, and our wealth at time n is

$$W_n = W_0 + \sum_{m=1}^{n} H_m \xi_m$$

Introducing $Y_n = Y_0 + \xi_1 + \cdots + \xi_n$ as the wealth of a gambler who bets 1 unit every time, we can write the last formula as

(⋆) $$W_n = W_0 + \sum_{m=1}^{n} H_m (Y_m - Y_{m-1})$$

The last bit of arithmetic may look strange, but we can make the right-hand side of (⋆) sound more natural by considering Y_m to be the price of a stock at time m, and H_m the amount of stock we hold from time $(m-1)$ to time m. In this case our profit from time $m-1$ to m is the amount we hold times the change in the price of the stock: $H_m(Y_m - Y_{m-1})$. Summing up our profits and adding our initial wealth then gives our wealth at time n.

We are now able to state the theoretical result that implies (3.1).

(3.2) Theorem. *Suppose that Y_n is a supermartingale with respect to X_n and that $H_n \in \mathcal{F}(X_0, \ldots, X_{n-1})$ satisfies $0 \le H_n \le c_n$ where c_n is a constant that may depend on n. Then*

$$W_n = W_0 + \sum_{m=1}^{n} H_m(Y_m - Y_{m-1}) \quad \text{is a supermartingale}$$

We need the condition $H_n \ge 0$ to prevent the bettor from becoming the house by betting a negative amount of money. The upper bound $H_n \le c_n$ is a technical condition that is needed to have expected values make sense. In the gambling context this assumption is harmless: even if the bettor wins every time there is an upper bound to the amount of money he can have at time n.

Proof. The gain at time $n+1$ is

$$W_{n+1} - W_n = H_{n+1}(Y_{n+1} - Y_n)$$

Let $A = \{X_n = i_n, X_{n-1} = i_{n-1}, \ldots, X_0 = i_0\}$. Since H_{n+1} is a function of $X_0, X_1, \ldots, X_n$, it is constant on the event A, and (1.3) implies

$$E(H_{n+1}(Y_{n+1} - Y_n)|A) = H_{n+1}E(Y_{n+1} - Y_n|A) \le 0$$

verifying that W_n is a supermartingale. $\square$

We can now give the

Proof of (3.1). By (3.2) the winnings from the doubling strategy are a supermartingale. (2.3) then implies that $EW_n \le EW_0 = 0$. $\square$

Though (3.2) may be depressing for gamblers, a simple special case gives us an important computational tool. To introduce this tool, we need one more notion.

Definition. We say that T is a **stopping time with respect to** X_n if the occurrence (or nonoccurrence) of the event "we stop at time n" can be determined by looking at the values of the process up to that time: $X_0, X_1, \ldots, X_n$.

Example 3.2. Constant betting up to a stopping time. One possible gambling strategy is to bet \$1 each time until you stop playing at time T. In symbols, we let $H_m = 1$ if $T > m$ and 0 otherwise. To check that this is an admissible gambling strategy we note that the set on which H_m is 0 is

$$\{T > m\}^c = \{T \le m - 1\} = \cup_{k=1}^{m-1}\{T = k\}$$

By the definition of a stopping time, the event $\{T = k\}$ can be determined from the values of $X_0, \ldots, X_k$. Since the union is over $k \le m - 1$, H_m can be determined from the values of $X_0, X_1, \ldots, X_{m-1}$.

Having introduced the gambling strategy "Bet \$1 on each play up to time T" our next step is to compute the payoff we receive. Letting $T \wedge n$ denote the minimum of T and n, i.e., it is T if $T < n$ and n otherwise we can give the answer as:

(♯) $$W_n = Y_0 + \sum_{m=1}^{n} H_m(Y_m - Y_{m-1}) = Y_{T \wedge n}$$

To check the last equality, consider two cases:

(i) if $T > n$ then $H_m = 1$ for all $m \le n$, so

$$W_n = Y_0 + (Y_n - Y_0) = Y_n$$

(ii) if $T \le n$ then $H_m = 0$ for $m > T$ and the sum stops at T. In this case,

$$W_n = Y_0 + (Y_T - Y_0) = Y_T$$

which again agrees with the right-hand side of (♯).

Combining (♯) with (3.2), shows that $W_n = Y_{T \wedge n}$ is a supermartingale so we have:

(3.3) Theorem. *If Y_n is a supermartingale with respect to X_n and T is a stopping time with respect to X_n then the stopped process $Y_{T \wedge n}$ is a supermartingale with respect to X_n.*

Multiplying the last result by -1 we can see:

(3.4) Theorem. *If Y_n is a submartingale with respect to X_n and T is a stopping time with respect to X_n, then the stopped process $Y_{T \wedge n}$ is a submartingale with respect to X_n.*

Since a process is a martingale if and only if it is both a submartingale and a supermartingale, it follows that:

(3.5) **Theorem.** *If M_n is a martingale with respect to X_n and T is a stopping time with respect to X_n, then the stopped process $M_{T \wedge n}$ is a martingale with respect to X_n.*

The next example shows that if there are no restrictions on the stopping time, then we can make money by stopping a martingale.

Example 3.3. Waiting for a dollar profit. Let $\xi_1, \xi_2, \ldots$ be independent and have $P(\xi_i = 1) = 1/2$, $P(\xi_i = -1) = 1/2$. Let $S_n = \xi_1 + \xi_2 + \cdots + \xi_n$, and let $T_1 = \min\{n \geq 1 : S_n = 1\}$. Results in Example 7.1 of Chapter 1 imply that $P(T_1 < \infty) = 1$. At time T_1 we have $S_{T_1} = 1$, i.e., a sure profit of 1. The next result says that this cannot happen unless we have an unlimited amount of credit.

(3.6) **Stopping theorem for bounded martingales.** *Let M_n be a martingale with respect to X_n, and let T be a stopping time with respect to X_n, with $P(T < \infty) = 1$. If there is a constant K so that $|M_{T \wedge n}| \leq K$ for all n, then*

$$EM_T = EM_0$$

Proof. (3.5) and (2.5) imply that $EM_0 = EM_{T \wedge n}$. Using this and the condition $|M_{T \wedge n}| \leq K$, it follows that

$$|EM_0 - EM_T| = |EM_{T \wedge n} - EM_T| \leq 2KP(T > n) \to 0$$

as $n \to \infty$. So we must have $|EM_0 - EM_T| = 0$. That is, $EM_0 = EM_T$. □

The next section is devoted to applications and extensions of this result.

2.4. Applications

Our first two examples are from Section 1.6.

Example 4.1. Simple random walk. Let $S_n = S_0 + \xi_1 + \cdots + \xi_n$ where $\xi_1, \xi_2, \ldots \xi_n$ are independent and have $P(\xi_i = 1) = P(\xi_i = -1) = 1/2$. It follows from Example 2.1 that S_n is a martingale. Let $T = \min\{n : S_n \notin (a,b)\}$ be the first time S_n leaves the interval (a,b). To check that T is a stopping time with respect to S_n we note that

$$\{T > n\} = \{S_0, S_1, \ldots, S_n \in (a,b)\}$$

which can be determined by looking at $S_0, \ldots, S_n$.

The fact that $P(T < \infty) = 1$ is a little more tricky, but can be shown by noting that no matter where we are in the interval (a, b), a sequence $b - a$ consecutive up jumps will drive us out, so the pedestrian lemma, (3.3) in Chapter 1, implies that this will happen eventually. Since $S_{T \wedge n} \in (a, b)$, using (3.6) now shows that if $S_0 = x$, then

$$x = ES_T = aP(S_T = a) + bP(S_T = b)$$

Since $P(S_T = a) = 1 - P(S_T = b)$ solving gives $x = a + (b-a)P(S_T = b)$. Introducing $V_y = \min\{n \geq 0 : X_n = y\}$, we have

$$(4.1) \qquad P_x(V_b < V_a) = \frac{x-a}{b-a}$$

generalizing (6.2) in Chapter 1.

EXERCISE 4.1. Apply this reasoning to the Wright–Fisher model with no mutation (Example 6.5 in Chapter 1) to conclude that $P_x(V_N < V_0) = x/N$.

Example 4.2. Gambler's ruin. This time let $S_n = S_0 + \xi_1 + \cdots + \xi_n$ where $\xi_1, \xi_2, \ldots \xi_n$ are independent with $P(\xi_i = 1) = p$ and $P(\xi_i = -1) = q$ where $q = 1 - p$. Suppose $0 < p < 1$ and let $g(x) = (q/p)^x$. Example 2.6 implies that $g(S_n)$ is a martingale. Let $T = \min\{n : S_n \notin (a, b)\}$. The arguments in the previous example show that T is a stopping time and $P(T < \infty) = 1$. Since $g(S_{T \wedge n})$ is bounded, it follows from (3.10) that if $S_0 = x$

$$(q/p)^x = (q/p)^a P(S_T = a) + (q/p)^b P(S_T = b)$$

Since $P(S_T = a) = 1 - P(S_T = b)$ we have

$$(q/p)^x = (q/p)^a + \{(q/p)^b - (q/p)^a\} P(S_T = b)$$

Solving gives

$$(4.2) \qquad P_x(V_b < V_a) = \frac{(q/p)^x - (q/p)^a}{(q/p)^b - (q/p)^a}$$

generalizing (6.2) from Chapter 1. □

Consider now a random walk $S_n = S_0 + \xi_1 + \cdots + \xi_n$ where $\xi_1, \xi_2, \ldots$ are independent and have the same distribution. We will now derive three formulas by using three martingales: one linear, one quadratic, and one exponential in S_n. Assume that the mean exists, i.e., $E|\xi_i| < \infty$, and let $\mu = E\xi_i$ be the mean movement on one step. Remarks after Example 2.1 imply that $M_n = S_n - n\mu$ is a martingale with respect to S_n. Combining this observation with the stopping theorem, (3.6), leads us to:

(4.3) Wald's equation. *Let S_n be a random walk. If T is a stopping time with $ET < \infty$, then*
$$E(S_T - S_0) = \mu ET$$

Why is this true? The stopping theorem (3.6), and the fact that martingale expected values are constant, (3.5), imply that

(4.4) $$E(S_{T \wedge n} - \mu E(T \wedge n)) = ES_0$$

Under the assumptions imposed, one can let $n \to \infty$ in the last equation and rearrange to conclude $E(S_T - S_0) = \mu ET$. □

With Wald's equation in hand we can now better understand:

Example 4.3. Mean time to gambler's ruin. As in the previous example, $S_n = S_0 + \xi_1 + \cdots + \xi_n$ where $\xi_1, \xi_2, \ldots \xi_n$ are independent with $P(\xi_i = 1) = p$ and $P(\xi_i = -1) = 1 - p$. Let $V_a = \min\{n \geq 0 : S_n = a\}$. The mean movement on one step is $\mu = 2p - 1$, so using a subscript x to indicate $S_0 = x$, Wald's equation implies that if $p < 1/2$ and $x > 0$, then

(4.5) $$E_x V_0 = x/(1 - 2p)$$

Why is this true? We lose an average of $(1 - 2p)$ dollars per play, so it should take us an average of $x/(1-2p)$ plays to lose our initial $\$x$. To extract this from Wald's equation, let $T = V_0$ and note that $S_0 = x$ while $S_T = 0$, so (4.3) tells us that $-x = \mu E_x V_0$. Since $\mu = -(1 - 2p)$, it follows that $x = (1 - 2p)E_x V_0$. □

Proof. Alert readers have noticed that we did not prove that $E_x V_0 < \infty$ when $p < 1/2$. To do this we note that (4.4) implies
$$(1 - 2p)E_x(V_0 \wedge n) = E_x S_0 - E_x S_{V_0 \wedge n} \leq x$$
since $S_0 = x$ and $S_{V_0 \wedge n} \geq 0$. Letting $n \to \infty$, we have $E_x V_0 \leq x/(1 - 2p) < \infty$ and the application of Wald's equation is legitimate. □

EXERCISE 4.2. In order to reach 0 we must first go from x to $x - 1$, then from $x - 1$ to $x - 2$, and so on. Use this observation to argue that the answer must be of the form $E_x T_0 = cx$.

EXERCISE 4.3. Use (4.3) and an argument by contradiction to show that if $p \geq 1/2$, then $E_x V_0 = \infty$ for $x > 0$.

As we mentioned earlier the martingale of Example 1.5 allows us to derive the result in Example 6.5 of Chapter 1.

Example 4.4. Duration of fair games. Consider a symmetric simple random walk S_n with $S_0 = 0$ and let $\tau = \min\{n : S_n \notin (a,b)\}$ where $a < 0 < b$. To explain our interest in this problem consider a gambler with a dollars who is playing a fair game and will quit when he has b dollars. In this case $x_i = 1$ and $x = -1$ with probability $1/2$ each so the variance $\sigma^2 = Ex_i^2 = 1$. We claim that

(4.6) $$E_0 \tau = -ab$$

Why is this true? Ignoring the fact that $S_{\tau \wedge n}^2 - (\tau \wedge n)$ is not a bounded martingale, we can apply (3.6) to conclude that $0 = E_0 S_\tau^2 - E_0 \tau$. Using (4.1), now we see that

$$E\tau = ES_\tau^2 = a^2 \cdot \frac{b}{b-a} + b^2 \cdot \frac{-a}{b-a} = \frac{-ab^2 + ba^2}{b-a} = -ab$$

Proof. To justify the last computation we let N be a positive integer and consider the stopping time $T = \tau \wedge N$, which has $|M_{T \wedge n}| \leq \max\{b^2, a^2, N\}$ for all n since $T \wedge n = T \wedge N$ for $n \geq N$. Using (3.6) now as before and rearranging, we have

$$E_0(\tau \wedge N) = ES_{\tau \wedge N}^2$$

and letting $N \to \infty$ gives (4.6). □

To see what (4.6) says we will now consider a concrete example. If we set $a = -r$ and $b = r$, then (4.6) implies that $E\tau_{-r,r} = r^2$. In particular when $r = 25$, we have $E\tau_{-25,25} = 625$. If we double the value of r to $r = 50$, we have $E\tau_{-50,50} = 2500$.

Our next topic is an application of the exponential martingale $M_n = \exp(\theta S_n)/\varphi(\theta)^n$ of Example 2.7.

Example 4.5. Left continuous random walk. Generalizing our analysis of the gambler's ruin chain in Example 4.2, we introduce the assumption:

(A) Suppose that $\xi_1, \xi_2, \ldots$ are independent integer-valued random variables with $E\xi_i > 0$, $P(\xi_i \geq -1) = 1$, and $P(\xi_i = -1) > 0$.

These walks are called left continuous since they cannot jump over any integers when they are decreasing, which is going to the left as the number line is usually drawn. Define $\alpha < 0$ by the requirement that $\varphi(\alpha) = 1$. To see that such an α exists, note that (i) $\varphi(0) = 1$ and

$$\varphi'(\theta) = \frac{d}{d\theta} E e^{\theta x_i} = E(x_i e^{\theta x_i}) \quad \text{so} \quad \varphi'(0) = E x_i > 0$$

and $\varphi(\theta) < 1$ for small negative θ. (ii) If $\theta < 0$, then $\varphi(\theta) \geq e^{-\theta} P(x_i = -1) \to \infty$ as $\theta \to -\infty$. Our choice of α makes $\exp(\alpha S_n)$ a martingale. Having found the martingale it is easy now to conclude:

(4.7) Let $a < 0$ and $T_a = \min\{n : S_n = a\}$. Under assumption (A), we have

$$P_0(T_a < \infty) = e^{-\alpha a}$$

Why is this true? Ignoring the fact that $P_0(V_a < \infty) < 1$ and trusting that there will be no contribution to the expectation from $\{V_a = \infty\}$ we can use (3.6) to conclude

$$1 = E_0 \exp(\alpha S_{V_a}) = e^{\alpha a} P_0(V_a < \infty)$$

Solving gives the desired result. □

Proof. To cope with the fact that $P_0(V_a < \infty) < 1$, we apply (3.6) to $T = V_a \wedge n$ to conclude that

$$1 = E_0 \exp(\alpha S_{V_a \wedge n}) = e^{\alpha a} P_0(V_a \leq n) + E_0(\exp(\alpha S_n); T > n)$$

The strong law of large numbers implies that on $T = \infty$, $S_n/n \to \mu > 0$, so the second term $\to 0$ as $n \to \infty$ and it follows that $1 = e^{\alpha a} P_0(V_a < \infty)$. □

Example 4.6. Ruin in a favorable game. Consider the special case in which $P(\xi_i = 1) = p > 1/2$, and $P(\xi_i = -1) = q = 1 - p$. As noted in Example 2.7,

$$\varphi(\theta) = E e^{\theta X_i} = p e^\theta + q e^{-\theta}$$

so if we pick $e^\alpha = q/p$ then $\varphi(\alpha) = 1$. Using (4.7) now, it follows that if $a < 0$ then $P_0(T_a < \infty) = (q/p)^{-a}$. Changing variables $x = -a$ and using symmetry, we see that if $x > 0$, then

(4.8) $$P_x(T_0 < \infty) = (q/p)^x$$

Section 2.4 Applications 119

When the random walk is not left continuous we cannot get exact results on hitting probabilities but we can still get a bound.

Example 4.7. Cramer's estimate of ruin. Let S_n be the total assets of an insurance company at the end of year n. During year n, premiums totaling c dollars are received, while claims totaling ζ_n dollars are paid, so

$$S_n = S_{n-1} + c - \zeta_n$$

Let $\eta_n = c - \zeta_n$ and suppose that $\eta_1, \eta_2, \ldots$ are independent random variables that are normal with mean $\mu > 0$ and variance σ^2. That is the density function of η_i is

$$(2\pi\sigma^2)^{-1/2} \exp(-(x-\mu)^2/2\sigma^2)$$

Let B for bankrupt be the event that the wealth of the insurance company is negative at some time n. We will show

(4.9) $$P(B) \leq \exp(-2\mu S_0/\sigma^2)$$

In words, in order to be successful with high probability, $\mu S_0/\sigma^2$ must be large, but the failure probability decreases exponentially fast as this quantity increases.

Proof. We begin by computing $\varphi(\theta) = E \exp(\theta \eta_i)$. To do this we need a little algebra

$$-\frac{(x-\mu)^2}{2\sigma^2} + \theta(x-\mu) + \theta\mu = -\frac{(x-\mu-\sigma^2\theta)^2}{2\sigma^2} + \frac{\sigma^2\theta^2}{2} + \theta\mu$$

and a little calculus

$$\varphi(\theta) = \int e^{\theta x} (2\pi\sigma^2)^{-1/2} \exp(-(x-\mu)^2/2\sigma^2) \, dx$$

$$= \exp(\sigma^2\theta^2/2 + \theta\mu) \int (2\pi\sigma^2)^{-1/2} \exp\left(-\frac{(x-\mu-\sigma^2\theta)^2}{2\sigma^2}\right) dx$$

Since the integrand is the density of a normal with mean $\mu + \sigma^2\theta$ and variance σ^2 it follows that

(4.10) $$\varphi(\theta) = \exp(\sigma^2\theta^2/2 + \theta\mu)$$

If we pick $\theta = -2\mu/\sigma^2$, then

$$\sigma^2\theta^2/2 + \theta\mu = 2\mu^2/\sigma^2 - 2\mu^2/\sigma^2 = 0$$

So (2.10) implies $\exp(-2\mu S_n/\sigma^2)$ is a martingale. Let $T = \min\{n : S_n \leq 0\}$. Applying (4.4) to $T \wedge n$ we conclude that

$$\exp(-2\mu S_0/\sigma^2) = E\exp(-2\mu S_{T\wedge n}) \geq P(T \leq n)$$

since $\exp(-2\mu S_T/\sigma^2) \geq 1$ and the contribution to the expected value from $\{T > n\}$ is ≥ 0. Letting $n \to \infty$ now and noticing $P(T \leq n) \to P(B)$ gives the desired result. □

Our final application uses the backwards random walk in Example 2.10.

Example 4.8. Ballot theorem. Suppose that in an election Al gets a votes while Betty gets b votes. Let $n = a + b$ be the total number of votes cast and suppose that Al wins, i.e., $a > b$.

(4.11) *The probability that Al always leads during the counting of the votes is $(a-b)/n$, i.e., his margin of victory divided by the total number of votes cast.*

Proof. Let $\xi_1, \xi_2, \ldots$ be independent and equal to 0 or 2 with probability 1/2 each and let $S_m = \xi_1 + \cdots + \xi_m$. Thinking of 2's as votes for Betty and 0's as votes for Al, the event that Al always leads in the counting is

$$G = \{S_j < j \text{ for all } j\}$$

We are interested in $P(G|S_n = 2b)$. To compute this probability we use our backwards martingale

$$M_0 = S_n/n, M_1 = S_{n-1}/(n-1), \ldots M_{n-2} = S_2/2, M_{n-1} = S_1/1$$

which starts at $M_0 = 2b/n < 1$. Let $T = \min\{m : M_m = 1 \text{ or } m = n-1\}$ In words, this is the first time m (working backwards from the end) that Al and Betty have the same number of votes or $m = n-1$ if no such time exists. On G^c, we have $M_T = 1$. In words, the first time Al is not ahead the number of votes must be equal. In contrast on G, Al is never behind, we have $T = n-1$ and $M_T = S_1/1 = 0$ since the first vote must be for Al if he always leads. Using the stopping theorem now gives

$$\frac{2b}{n} = M_0 = EM_T = P(G^c) \cdot 1 + P(G) \cdot 0$$

Thus $P(G^c) = 2b/n = 2b/(a+b)$ and

$$P(G) = 1 - \frac{2b}{a+b} = \frac{a-b}{a+b}$$

□

2.5. Exercises

Throughout the exercises we will use our standard notion for hitting times. $T_a = \min\{n \geq 1 : X_n = a\}$ and $V_a = \min\{n \geq 0 : X_n = a\}$.

5.1. *Brother–sister mating.* Consider the six state chain defined in Exercise 9.10 of Chapter 1. Show that the total number of A's is a martingale and use (3.6) to compute the probability of getting absorbed into the 2,2 (i.e., all A's state) starting from each initial state.

5.2. *Martingale Markov chains.* Let X_n be a Markov chain with state space $\{0, 1, \ldots, N\}$ and suppose that X_n is a martingale. (a) Show that 0 and N must be absorbing states. (b) Let $\tau = V_0 \wedge V_N$. Suppose $P_x(\tau < \infty) > 0$ for $1 < x < N$. Show that $P_x(\tau < \infty) = 1$ and $P_x(V_N < V_0) = x/N$.

5.3. *Lognormal stock prices.* Consider the special case of Example 2.3 in which $\zeta_i = e^{\eta_i}$ where $\eta_i = \text{normal}(\mu, \sigma^2)$. For what values of μ and σ is $X_n = X_0 \cdot \zeta_1 \cdots \zeta_n$ a martingale?

5.4. *Likelihood ratio for normals.* Take $f = \text{normal}(\mu, 1)$, $g = \text{normal}(0, 1)$ in Example 2.8. (a) Compute the martingale that results in (2.11). (b) Let $S_n = \xi_1 + \cdots + \xi_n$ and show that the martingale you found in (a) is a special case of the exponential martingale in Example 2.7.

5.5. *An unfair fair game.* Define random variables recursively by $X_0 = 1$ and for $n \geq 1$, X_n is chosen uniformly on $(0, X_{n-1})$. If we let $U_1, U_2, \ldots$ be uniform on $(0, 1)$, then we can write this sequence as $X_n = U_n U_{n-1} \cdots U_0$. (a) Show that $M_n = 2^n X_n$ is a martingale. (b) Use the fact that $\log X_n = \log U_1 + \cdots + \log U_n$ to show that $(1/n) \log X_n \to -1$. (c) Use (b) to conclude $M_n \to 0$, i.e., in this "fair" game our fortune always converges to 0 as time tends to ∞.

5.6. Let X_n be the Wright–Fisher model with no mutation discussed in Example 2.4. (a) Show that $Y_n = X_n(N - X_n)/(1 - 1/N)^n$ is a martingale. (b) Use this to conclude that

$$(N-1) \leq \frac{x(N-x)(1-1/N)^n}{P_x(0 < X_n < N)} \leq \frac{N^2}{4}$$

5.7. Consider the urn scheme of Example 2.10. At time 0 there is one ball of each color. At time n we draw out a ball chosen at random. We return it to the urn and add one more ball of the color chosen. (a) Compute the probability that the first k draws are red and the next $n - k$ are green. (b) Generalize from this to conclude that if X_n is the fraction of red balls at time n then $P(X_n = k/(n+2)) = 1/n + 1$ for $1 \leq k \leq n+1$.

122   Chapter 2   Martingales

5.8. Modify the set-up in the previous problem so that we start with 5 green and 1 red. Use the stopping theorem, (3.6), to find an upper bound on the probability of $A = \{X_n > 1/2 \text{ for some } n \geq 0\}$.

5.9. Let $X_n \geq 0$ be a supermartingale and $\lambda > 0$. Show that

$$P\left(\max_{n \geq 0} X_n > \lambda\right) \leq EX_0/\lambda$$

5.10. Let $X_n \geq 0$ be a submartingale. (a) Show that if T is a stopping time with $P(T \leq n) = 1$, then $EX_T \leq EX_n$. (b) Use (a) to show that if $\lambda > 0$, then

$$P\left(\max_{0 \leq m \leq n} X_m > \lambda\right) \leq EX_n/\lambda$$

5.11. Let $S_n = \xi_1 + \cdots + \xi_n$ where the ξ_i are independent with $E\xi_i = 0$ and $\text{var}(X_i) = \sigma^2$. Use the previous exercise to show that

$$P\left(\max_{0 \leq m \leq n} |S_m| > \rho\right) \leq n\sigma^2/\rho^2$$

5.12. Consider a favorable game in which the payoffs are -1, 1, or 2 with probability 1/3 each. Use the results of Example 4.5 to compute the probability we ever go broke (i.e, our winnings W_n reach \$0) when we start with \$$i$.

5.13. *Duration of unfair games.* Let $\xi_1, \xi_2, \ldots$ be independent with $P(\xi_i = 1) = p$ and $P(\xi_i = -1) = q = 1 - p$ where $p \neq 1/2$. Let $S_n = S_0 + \xi_1 + \cdots + \xi_n$, and let $\tau = \min\{n : S_n \notin (0, N)\}$. Combine Wald's equation and the exit formula in (4.2) to show that

$$E_x \tau = \frac{x}{q-p} - \frac{N}{q-p} \cdot \frac{1-(q/p)^x}{1-(q/p)^N}$$

giving another solution of Exercise 6.1 in Chapter 1.

5.14. *Variance of the time of gambler's ruin.* Let $\xi_1, \xi_2, \ldots$ be independent with $P(\xi_i = 1) = p$ and $P(\xi_i = -1) = q = 1 - p$ where $p < 1/2$. Let $S_n = S_0 + \xi_1 + \cdots + \xi_n$. In Example 4.3 we showed that if $V_0 = \min\{n \geq 0 : S_n = 0\}$ then $E_x V_0 = x/(1-2p)$. The aim of this problem is to compute the variance of V_0. (a) Show that $(S_n - (p-q)n)^2 - n(1-(p-q)^2)$ is a martingale. (b) Use this to conclude that when $S_0 = x$ the variance of V_0 is

$$x \cdot \frac{1-(p-q)^2}{(p-q)^3}$$

(c) Why must the answer in (b) be of the form cx?

5.15. Generating function of the time of gambler's ruin. Continue with the set-up of the previous problem. (a) Use the exponential martingale and our stopping theorem to conclude that if $\theta \le 0$, then $e^{\theta x} = E_x(\varphi(\theta)^{-V_0})$. (b) Let $0 < s < 1$. Solve the equation $\varphi(\theta) = 1/s$, then use (a) to conclude

$$E_x(s^{V_0}) = \left(\frac{1 - \sqrt{1 - 4pqs^2}}{2ps} \right)^x$$

(c) Why must the answer in (b) be of the form $f(s)^x$?

5.16. General birth and death chains. The state space is $\{0, 1, 2, \ldots\}$ and the transition probability has

$$p(x, x+1) = p_x$$
$$p(x, x-1) = q_x \quad \text{for } x > 0$$
$$p(x, x) = r_x \quad \text{for } x \ge 0$$

while the other $p(x, y) = 0$. Let $V_y = \min\{n \ge 0 : X_n = y\}$ be the time of the first visit to y and let $h_N(x) = P_x(V_N < V_0)$. Let $\varphi(z) = \sum_{y=1}^{z} \prod_{x=1}^{y-1} q_x/p_x$. Show that

$$P_x(V_b < V_a) = \frac{\varphi(x) - \varphi(a)}{\varphi(b) - \varphi(a)}$$

From this it follows that 0 is recurrent if and only if $\varphi(b) \to \infty$ as $b \to \infty$, giving another solution of Exercise 9.46 from Chapter 1.

5.17. Lyapunov functions. Let X_n be an irreducible Markov chain with state space $\{0, 1, 2, \ldots\}$ and let $\varphi \ge 0$ be a function with $\lim_{x \to \infty} \varphi(x) = \infty$, and $E_x\varphi(X_1) \le \varphi(x)$ when $x \ge K$. Then X_n is recurrent. This abstract result is often useful for proving recurrence in many chains that come up in applications and in many cases it is enough to consider $\varphi(x) = x$.

5.18. $GI/G/1$ queue. Let $\xi_1, \xi_2, \ldots$ be independent with distribution F and Let $\eta_1, \eta_2, \ldots$ be independent with distribution G. Define a Markov chain by

$$X_{n+1} = (X_n + \xi_n - \eta_{n+1})^+$$

where $y^+ = \max\{y, 0\}$. Here X_n is the workload in the queue at the time of arrival of the nth customer, not counting the service time of the nth customer, η_n. The amount of work in front of the $(n+1)$th customer is that in front of the nth customer plus his service time, minus the time between the arrival of customers n and $n+1$. If this is negative the server has caught up and the waiting time is 0. Suppose $E\xi_i < E\eta_i$ and let $\epsilon = (E\eta_i - E\xi_i)/2$. (a) Show that there is a K so that $E_x(X_1 - x) \le -\epsilon$ for $x \ge K$. (c) Let $U_k = \min\{n : X_n \le$

K}. (b) Use the fact that $X_{U_k \wedge n} + \epsilon(U_k \wedge n)$ is a supermartingale to conclude that $E_x U_k \le x/\epsilon$.

5.19. *Hitting probabilities.* Consider a Markov chain with finite state space S. Let a and b be two points in S that are absorbing states, let $\tau = V_a \wedge V_b$, and let $C = S - \{a, b\}$. Suppose $h(a) = 1$, $h(b) = 0$, and for $x \in C$ we have

$$h(x) = \sum_y p(x, y) h(y)$$

(a) Show that $h(X_n)$ is a martingale. (b) Use (3.8) to show that if we have $P_x(\tau < \infty) > 0$ for all $x \in C$, then $h(x) = P_x(V_a < V_b)$.

5.20. *Expectations of hitting times.* Consider a Markov chain with finite state space S. Let $A \subset S$ be a closed set, let $V_A = \min\{n \ge 0 : X_n \in A\}$ be the time of the first visit to A. Suppose that $g(x) = 0$ for $x \in A$, while for $x \in B = S - A$ we have

$$g(x) = 1 + \sum_y p(x, y) g(y)$$

(a) Show that $g(X_{V_A \wedge n}) - (V_A \wedge n)$ is a martingale. (b) Use (3.6) to show that if $P_x(V_A < \infty) > 0$ for all $x \in C$ then $g(x) = E_x V_A$.

5.21. *Second moments of branching processes.* Consider the branching process in Example 2.2 and suppose that $\sum_k k^2 p_k < \infty$. (a) Show that

$$E(Z_n^2 | Z_{n-1} = z) = \sigma^2 Z_{n-1} + (\mu Z_{n-1})^2$$

(b) Take expected values in (a) to conclude that if $X_n = Z_n/\mu^n$ and $Z_0 = 1$,

$$EX_n^2 = EX_{n-1}^2 + \sigma^2/\mu^{n+1}$$

and hence $EX_n^2 = 1 + \sigma^2 \sum_{k=2}^{n+1} \mu^{-k}$. Note that this is $n\sigma^2 + 1$ if $\mu = 1$. The second moment stays bounded if $\mu > 1$, but blows up when $\mu < 1$.

5.22. *Critical branching processes.* Consider the branching process in Example 2.2 and suppose that the mean number of offspring is $\mu = 1$. Exclude the trivial case by supposing $p_1 < 1$, and simplify the problem by supposing that $p_k = 0$ if $k > K$. Let K be a large number and let $U_K = \min\{n : Z_n > K\}$. (a) Use the pedestrian lemma to conclude that if $Z_n \le K$ for all n, then Z_n will eventually hit zero. (b) Use (3.6) to conclude that $P_x(U_K < \infty) \le x/K$. (c) Combine (a) and (b) to conclude that $P_x(V_0 < \infty) = 1$.

5.23. *Improved stopping theorem.* Let M_n be a martingale with respect to X_n, and let T be a stopping time with respect to X_n with $P(T < \infty) = 1$. Suppose that $E|M_T| < \infty$ and $\lim_{n\to\infty} E(M_n; T > n) = 0$. Use the proof of (3.6) to conclude that $EM_T = EM_0$.

5.24. *Martingales with bounded jumps.* Let M_n be a martingale with respect to X_n, with $M_0 = 0$ and suppose $|M_{n+1} - M_n| \le K$ for all n. Let T be a stopping time with $ET < \infty$. Use the result in the previous exercise to conclude that $EM_T = 0$.

5.25. *Random walks with bounded jumps.* Let $\xi_1, \xi_2, \ldots$ be independent and have the same distribution, which we assume has $E\xi_i = 0$ and $|\xi_i| \le K$. Let $S_n = \xi_1 + \cdots + \xi_n$ and $V_x = \min\{n \ge 0 : S_n = x\}$. Use the previous exercise to show that if $x \ne 0$, then $E_0 V_x = \infty$.

3 Poisson Processes

3.1. Exponential Distribution

To prepare for our discussion of the Poisson process, we need to recall the definition and some of the basic properties of the exponential distribution. A random variable T is said to have **an exponential distribution with rate** λ, or $T = \text{exponential}(\lambda)$, if

$$(1.1) \qquad P(T \leq t) = 1 - e^{-\lambda t} \quad \text{for all } t \geq 0$$

Here we have described the distribution by giving the **distribution function** $F(t) = P(T \leq t)$. We can also write the definition in terms of the **density function** $f_T(t)$ which is the derivative of the distribution function.

$$(1.2) \qquad f_T(t) = \begin{cases} \lambda e^{-\lambda t} & \text{for } t \geq 0 \\ 0 & \text{for } t < 0 \end{cases}$$

For the purposes of understanding and remembering formulas, it is useful to think of $f_T(t)$ as "$P(T = t)$," even though the probability that T is exactly equal to your favorite value of t is 0. We will use this notation for the density function in what follows. People who do not like this notation can simply substitute $f_T(t)$ for $P(T = t)$ to obtain a more "formally correct" but often less intuitive calculation.

Integrating by parts with $f(t) = t$ and $g'(t) = \lambda e^{-\lambda t}$,

$$(1.3) \qquad \begin{aligned} ET &= \int t\, P(T = t)\, dt = \int_0^\infty t \cdot \lambda e^{-\lambda t}\, dt \\ &= -t e^{-\lambda t} \Big|_0^\infty + \int_0^\infty e^{-\lambda t}\, dt = 1/\lambda \end{aligned}$$

Integrating by parts with $f(t) = t^2$ and $g'(t) = \lambda e^{-\lambda t}$, we see that

$$ET^2 = \int t^2\, P(T=t)\, dt = \int_0^\infty t^2 \cdot \lambda e^{-\lambda t}\, dt$$
$$= -t^2 e^{-\lambda t}\Big|_0^\infty + \int_0^\infty 2t e^{-\lambda t}\, dt = 2/\lambda^2$$

by the formula for ET. So the variance

(1.4) $$\text{var}(T) = ET^2 - (ET)^2 = 1/\lambda^2$$

While calculus is required to know the exact values of the mean and variance, it is easy to see how they depend on λ. Let $T = \text{exponential}(\lambda)$, i.e., have an exponential distribution with rate λ, and let $S = \text{exponential}(1)$. To see that S/λ has the same distribution as T, we use (1.1) to conclude

$$P(S/\lambda \leq t) = P(S \leq \lambda t) = 1 - e^{-\lambda t} = P(T \leq t)$$

Recalling that if c is any number then $E(cX) = cEX$ and $\text{var}(cX) = c^2 \text{var}(X)$, we see that

$$ET = ES/\lambda \qquad \text{var}(T) = \text{var}(S)/\lambda^2$$

Lack of memory property. It is traditional to formulate this property in terms of waiting for an unreliable bus driver. In words, "if we've been waiting for t units of time then the probability we must wait s more units of time is the same as if we haven't waited at all." In symbols

(1.5) $$P(T > t + s | T > t) = P(T > s)$$

To prove this we recall that if $B \subset A$, then $P(B|A) = P(B)/P(A)$, so

$$P(T > t + s | T > t) = \frac{P(T > t + s)}{P(T > t)} = \frac{e^{-\lambda(t+s)}}{e^{-\lambda t}} = e^{-\lambda s} = P(T > s)$$

where in the third step we have used the fact $e^{a+b} = e^a e^b$.

Exponential races. Let $S = \text{exponential}(\lambda)$ and $T = \text{exponential}(\mu)$ be independent. In order for the minimum of S and T to be larger than t, each of S and T must be larger than t. Using this and independence we have

(1.6) $$P(\min(S,T) > t) = P(S > t, T > t)$$
$$= P(S > t) P(T > t) = e^{-\lambda t} e^{-\mu t} = e^{-(\lambda+\mu)t}$$

That is, $\min(S,T)$ has an exponential distribution with rate $\lambda + \mu$. The last calculation extends easily to a sequence of independent random variables $T_1, \ldots, T_n$ where $T_i = \text{exponential}(\lambda_i)$.

(1.7)
$$P(\min(T_1, \ldots, T_n) > t) = P(T_1 > t, \ldots T_n > t)$$
$$= \prod_{i=1}^{n} P(T_i > t) = \prod_{i=1}^{n} e^{-\lambda_i t} = e^{-(\lambda_1 + \cdots + \lambda_n)t}$$

That is, the minimum, $\min(T_1, \ldots, T_n)$, of several independent exponentials has an exponential distribution with rate equal to the sum of the rates $\lambda_1 + \cdots \lambda_n$.

In the last paragraph we have computed the duration of a race between exponentially distributed random variables. We will now consider: "Who finishes first?" Going back to the case of two random variables, we break things down according to the value of S and then using independence with our formulas (1.1) and (1.2) for the distribution and density functions, to conclude

(1.8)
$$P(S < T) = \int_0^\infty P(S = s) P(T > s | S = s) \, ds$$
$$= \int_0^\infty \lambda e^{-\lambda s} e^{-\mu s} \, ds$$
$$= \frac{\lambda}{\lambda + \mu} \int_0^\infty (\lambda + \mu) e^{-(\lambda+\mu)s} \, ds = \frac{\lambda}{\lambda + \mu}$$

where on the last line we have used the fact that $(\lambda + \mu)e^{-(\lambda+\mu)s}$ is a density function and hence must integrate to 1. Of course, one can also use calculus to evaluate the integral.

From the calculation for two random variables, you should be able to guess that if $T_1, \ldots, T_n$ are independent exponentials, then

(1.9)
$$P(T_i = \min(T_1, \ldots, T_n)) = \frac{\lambda_i}{\lambda_1 + \cdots + \lambda_n}$$

That is, the probability of i finishing first is proportional to its rate.

Proof. Let $S = T_i$ and U be the minimum of T_j, $j \neq i$. (1.7) implies that U is exponential with parameter
$$\mu = \sum_{j \neq i} = (\lambda_1 + \cdots + \lambda_n) - \lambda_i$$

so using the result for two random variables
$$P(T_i = \min(T_1, \ldots, T_n)) = P(S < U) = \frac{\lambda_i}{\lambda_i + \mu} = \frac{\lambda_i}{\lambda_1 + \cdots + \lambda_n} \qquad \square$$

Let I be the (random) index of the T_i that is smallest. In symbols,

$$P(I = i) = \frac{\lambda_i}{\lambda_1 + \cdots + \lambda_n}$$

You might think that the T_i's with larger rates might be more likely to win early. However,

(1.10) I and $V = \min\{T_1, \ldots T_n\}$ are independent.

Proof. To compute the joint probability density

$$P(I = i, V = t) = P(T_i = t, T_j > t \text{ for } j \neq i) = \lambda_i e^{-\lambda_i t} \cdot \prod_{j \neq i} e^{-\lambda_j t}$$

$$= \frac{\lambda_i}{\lambda_1 + \cdots + \lambda_n} \cdot (\lambda_1 + \cdots + \lambda_n) e^{-(\lambda_1 + \cdots + \lambda_n)t}$$

$$= P(I = i) \cdot P(V = t)$$

since V has an exponential$(\lambda_1 + \cdots + \lambda_n)$ distribution. □

Our final fact in this section concerns sums of exponentials. Changing notation slightly to prepare for what we will do in the next section.

(1.11) **Theorem.** Let $t_1, t_2, \ldots$ be independent exponential(λ). The sum $T_n = t_1 + \cdots + t_n$ has a gamma(n, λ) distribution. That is, the density function of T_n is given by

$$P(T_n = t) = \lambda e^{-\lambda t} \cdot \frac{(\lambda t)^{n-1}}{(n-1)!} \quad \text{for } t \geq 0; \qquad 0 \text{ otherwise}$$

Proof. The proof is by induction on n. When $n = 1$, $T_1 = t_1$ has an exponential(λ) distribution. Recalling that the 0th power of any positive number is 1, and by convention we set $0! = 1$, the formula reduces to

$$P(T_1 = t) = \lambda e^{-\lambda t}$$

and we have shown that our formula is correct for $n = 1$.

To do the inductive step suppose that the formula is true for n. The sum $T_{n+1} = T_n + t_{n+1}$, so breaking things down according to the value of T_n, and using the independence of T_n and t_{n+1}, we have

$$P(T_{n+1} = t) = \int_0^t P(T_n = s) P(t_{n+1} = t - s) \, ds$$

130   Chapter 3   Poisson Processes

Plugging the formula from (1.11) in for the first term and the exponential density in for the second and using the fact that $e^a e^b = e^{a+b}$ with $a = -\lambda s$ and $b = -\lambda(t-s)$ gives

$$\int_0^t \lambda e^{-\lambda s} \frac{(\lambda s)^{n-1}}{(n-1)!} \cdot \lambda e^{-\lambda(t-s)} \, ds = e^{-\lambda t} \lambda^{n+1} \int_0^t \frac{s^{n-1}}{(n-1)!} \, ds$$

$$= \lambda e^{-\lambda t} \frac{\lambda^n t^n}{n!}$$

3.2. Defining the Poisson Process

In this section we will give two definitions of the **Poisson process with rate** λ. The first, which will be our official definition, is nice because it allows us to construct the process easily. The second definition, which will be proved to be equivalent the first one in (2.4), is a little more sophisticated, but it generalizes more easily when we want to define Poisson processes that have nonconstant arrival rates or that take place in dimensions $d > 1$.

Definition. Let $t_1, t_2, \ldots$ be independent exponential(λ) random variables. Let $T_n = t_1 + \cdots + t_n$ for $n \geq 1$, $T_0 = 0$, and define $N(s) = \max\{n : T_n \leq s\}$.

We think of the t_n as times between arrivals of customers at a bank, so $T_n = t_1 + \cdots + t_n$ is the arrival time of the nth customer, and $N(s)$ is the number of arrivals by time s. To check the last interpretation, consider the following example:

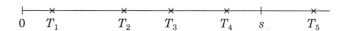

and note that $N(s) = 4$ when $T_4 \leq s < T_5$, that is, the 4th customer has arrived by time s but the 5th has not.

To explain why $N(s)$ is called the Poisson process rather than the exponential process, we will compute the distribution of $N(s)$. Now $N(s) = n$ if and only if $T_n \leq s < T_{n+1}$; i.e., the nth customer arrives before time s but the $(n+1)$th after s. Breaking things down according to the value of $T_n = t$ and noting that for $T_{n+1} > s$, we must have $t_{n+1} > s - t$, and t_{n+1} is independent

Section 3.2 Defining the Poisson Process

of T_n, it follows that

$$P(N(s) = n) = \int_0^s P(T_n = t)P(T_{n+1} > s | T_n = t)\, dt$$
$$= \int_0^s P(T_n = t)P(t_{n+1} > s - t)\, dt$$

Plugging in (1.11) now, the last expression is

$$= \int_0^s \lambda e^{-\lambda t} \frac{(\lambda t)^{n-1}}{(n-1)!} \cdot e^{-\lambda(s-t)}\, dt$$
$$= \frac{\lambda^n}{(n-1)!} e^{-\lambda s} \int_0^s t^{n-1}\, dt = e^{-\lambda s} \frac{(\lambda s)^n}{n!}$$

Recall that X has a **Poisson distribution** with mean μ, or $X = \text{Poisson}(\mu)$, for short, if

$$P(X = n) = e^{-\mu} \frac{\mu^n}{n!} \quad \text{for } n = 0, 1, 2, \ldots$$

Using our new definition, we can write the result of the last calculation as:

(2.1) Lemma. $N(s)$ has a Poisson distribution with mean λs.

This should explain the term "rate λ," the average number of arrivals in an amount of time s is λs, or λ per unit time.

Since this is our first mention of the Poisson distribution, we pause to let the reader derive some properties.

EXERCISE 2.1. (a) Show that if $X = \text{Poisson}(\mu)$ distribution, then

$$\varphi(s) \equiv Es^X = e^{-\mu(1-s)}$$

(b) Differentiate once and set $s = 1$ in (a) to conclude that $EX = \mu$.
(c) Differentiate k times and set $s = 1$ in (a) to conclude that

$$EX(X-1)\cdots(X-k+1) = \mu^k$$

(d) Use $\text{var}(X) = E(X(X-1)) + EX - (EX)^2$ to conclude $\text{var}(X) = \mu$. (e) Use (a) to conclude that if $X = \text{Poisson}(\mu)$ and $Y = \text{Poisson}(\nu)$ are independent, then $X + Y = \text{Poisson}(\mu + \nu)$.

The property of the Poisson process in (2.1) is the first part of our second definition. To start to develop the second part we show:

(2.2) **Lemma.** $N(t+s)-N(s)$, $t \geq 0$ is a rate λ Poisson process and independent of $N(r)$, $0 \leq r \leq s$.

Why is this true? Suppose for concreteness (and so that we can use the figure on page 130 again) that by time s there have been four arrivals T_1, T_2, T_3, T_4 which occurred at times u_1, u_2, u_3, u_4. We know that the waiting time for the fifth arrival must have $t_5 > s - u_4$, but by the lack of memory property of the exponential distribution

$$P(t_5 > s - u_4 + t | t_5 > s - u_4) = P(t_5 > t) = e^{-\lambda t}$$

This shows that the distribution of the first arrival after s is exponential(λ) and independent of T_1, T_2, T_3, T_4. It is clear that $t_5, t_6, \ldots$ are independent of T_1, T_2, T_3, T_4, and t_4. This shows that the interarrival times after s are independent exponential(λ), and hence that $N(t+s) - N(s)$, $t \geq 0$ is a Poisson process. □

From (2.2) we get easily the following:

(2.3) **Lemma.** $N(t)$ has **independent increments**: if $t_0 < t_1 < \ldots < t_n$, then

$$N(t_1) - N(t_0), N(t_2) - N(t_1), \ldots N(t_n) - N(t_{n-1}) \quad \text{are independent}$$

Why is this true? (2.2) implies that $N(t_n) - N(t_{n-1})$ is independent of $N(r)$, $r \leq t_{n-1}$ and hence of $N(t_{n-1}) - N(t_{n-2}), \ldots N(t_1) - N(t_0)$. The desired result now follows by induction. □

Combining (2.1) and (2.3) we have established half of:

(2.4) **Theorem.** If $\{N(s), s \geq 0\}$ is a Poisson process, then

(i) $N(0) = 0$,

(ii) $N(t + s) - N(s) = \text{Poisson}(\lambda t)$, and

(iii) $N(t)$ has independent increments.

Conversely, if (i), (ii), and (iii) hold, then $\{N(s), s \geq 0\}$ is a Poisson process.

Why is this true? As we remarked above, (2.1) and (2.3) give the first statement. To start to prove the converse, let T_n be the time of the nth arrival. The first arrival occurs after time t if and only if there were no arrivals in $[0, t]$. So using the formula for the Poisson distribution

$$P(T_1 > t) = P(N(t) = 0) = e^{-\lambda t}$$

Section 3.2 Defining the Poisson Process

This shows that $t_1 = T_1$ is exponential(λ). For $t_2 = T_2 - T_1$ we note that

$$\begin{aligned} P(t_2 > t | t_1 = s) &= P(\text{ no arrival in } (s, s+t] \, | t_1 = s) \\ &= P(N(t+s) - N(s) = 0 | N(r) = 0 \text{ for } r < s, N(s) = 1) \\ &= P(N(t+s) - N(s) = 0) = e^{-\lambda t} \end{aligned}$$

by the independent increments property in (iii), so t_2 is exponential(λ) and independent of t_1. Repeating this argument we see that $t_1, t_2, \ldots$ are independent exponential(λ). □

(2.4) is our second definition of the Poisson process. We said this definition is sophisticated because the existence of a process with these properties relies on the fact given in (e) of Exercise 2.1: if $X = \text{Poisson}(\mu)$ and $Y = \text{Poisson}(\nu)$ are independent, then $X + Y = \text{Poisson}(\mu + \nu)$. This is a very special property of the Poisson distribution. To illustrate this, suppose that we want to invent a new process in which increments have the (shifted) geometric distribution: $P(X = n) = (1-p)^n p$ for $n \geq 0$. Then we will fail since the sum of two independent shifted geometrics does not have a shifted geometric distribution.

$$\begin{aligned} P(X_1 + X_2 = k) &= \sum_{j=0}^{k} P(X_1 = j, X_2 = k - j) \\ &= \sum_{j=0}^{k} (1-p)^j p \cdot (1-p)^{k-j} p = k(1-p)^k p^2 \end{aligned}$$

Up to this point we have been concerned with the mechanics of defining the Poisson process, so the reader may be wondering:

Why is the Poisson process important for applications?

Our answer is based on the Poisson approximation to the binomial. Consider the Wendy's restaurant in the "Collegetown" section of Ithaca near Cornell. Suppose that each of the n individuals on the Cornell campus independently decides to go to Wendy's between 12:00 and 1:00 with probability λ/n, and suppose that a person who chooses to go, will do so at a time chosen at random from the interval. The probability that exactly k people will go is given by the binomial distribution

$$\frac{n(n-1)\cdots(n-k+1)}{k!} \left(\frac{\lambda}{n}\right)^k \left(1 - \frac{\lambda}{n}\right)^{n-k}$$

Exchanging the numerators of the first two fractions and breaking the last term into two, we have

$$(\star) \qquad \frac{\lambda^k}{k!} \cdot \frac{n(n-1)\cdots(n-k+1)}{n^k} \cdot \left(1 - \frac{\lambda}{n}\right)^n \left(1 - \frac{\lambda}{n}\right)^{-k}$$

Taking the four terms from left to right, we have

(i) $\lambda^k/k!$ does not depend on n.

(ii) There are k terms on the top and k terms on the bottom, so we can write this fraction as

$$\frac{n}{n} \cdot \frac{n-1}{n} \cdots \frac{n-k+1}{n}$$

The first term here is 1; the second is $1 - \frac{1}{n} \to 1$ as $n \to \infty$. This holds true for any fixed value of j, so the second term converges to 1 as $n \to \infty$.

(iii) It is one of the famous facts of calculus that $(1 - \lambda/n)^n \to e^{-\lambda}$ as $n \to \infty$. We have broken off the last term to be able to exactly apply this fact.

(iv) By the reasoning in (ii), $1 - \frac{\lambda}{n} \to 1$. The power $-k$ is fixed so

$$\left(1 - \frac{\lambda}{n}\right)^{-k} \to 1^{-k} = 1$$

Combining (i)–(iv), we see that $(\star)$ converges to

$$\frac{\lambda^k}{k!} \cdot 1 \cdot e^{-\lambda} \cdot 1$$

which is the Poisson distribution with mean λ.

By extending the last argument we can also see why the number of individuals that arrive in two disjoint time intervals should be independent. Using the multinomial instead of the binomial, we see that the probability j people will go between 12:00 and 12:20 and k people will go between 12:20 and 1:00 is

$$\frac{n!}{j!k!(n-j-k)!} \left(\frac{\lambda}{3n}\right)^j \left(\frac{2\lambda}{3n}\right)^k \left(1 - \frac{\lambda}{n}\right)^{n-(j+k)}$$

Rearranging gives

$$\frac{(\lambda/3)^j}{j!} \cdot \frac{(2\lambda/3)^k}{k!} \cdot \frac{n(n-1)\cdots(n-j-k+1)}{n^{j+k}} \cdot \left(1 - \frac{\lambda}{n}\right)^{n-(j+k)}$$

Reasoning as before shows that when n is large, this is approximately

$$\frac{(\lambda/3)^j}{j!} \cdot \frac{(2\lambda/3)^k}{k!} \cdot 1 \cdot e^{-\lambda}$$

Writing $e^{-\lambda} = e^{-\lambda/3} e^{-2\lambda/3}$ and rearranging we can write the last expression as

$$e^{-\lambda/3} \frac{(\lambda/3)^j}{j!} \cdot e^{-2\lambda/3} \frac{(2\lambda/3)^k}{k!}$$

This shows that the number of arrivals in the two time intervals are independent Poissons with means $\lambda/3$ and $2\lambda/3$.

The last proof can be easily generalized to show that if we divide the hour between 12:00 and 1:00 into any number of intervals, then the arrivals are independent Poissons with the right means. However, the argument gets very messy to write down.

More realistic models. Two of the weaknesses of the derivation above are:

(i) All individuals are assumed to have exactly the same probability of going to Wendy's.

(ii) Individuals who choose to go, do so at a time chosen at random between 12:00 and 1:00, so the arrival rate of customers is constant during the hour.

(i) is a very strong assumption but can be weakened by using a more general Poisson approximation result. For example,

(2.5) Theorem. *For each n let $X_{n,m}$ be independent random variables with $P(X_{n,m} = 1) = p_{n,m}$ and $P(X_{n,m} = 0) = 1 - p_{n,m}$. Let*

$$S_n = X_{n,1} + \cdots + X_{n,n}, \quad \lambda_n = ES_n = p_{n,1} + \cdots + p_{n,n},$$

and $Z_n = \text{Poisson}(\lambda_n)$. Then for any set A

$$|P(S_n \in A) - P(Z_n \in A)| \le \sum_{m=1}^{n} p_{n,m}^2$$

This result is nice since it gives a bound on the difference between the distribution of S_n and the Poisson distribution with mean $\lambda_n = ES_n$. Proofs of this can be found in a number of places. My favorite reference is Section 2.6 of Durrett (1995).

From (2.5) we immediately get the following convergence theorem.

136 Chapter 3 Poisson Processes

(2.6) Corollary. *Suppose that in (2.5), $\lambda_n \to \lambda < \infty$ and $\max_k p_{n,k} \to 0$, then*
$$\max_A |P(S_n \in A) - P(Z_n \in A)| \to 0$$

Proof. Since $p_{n,m}^2 \leq p_{n,m} \cdot \max_k p_{n,k}$, summing over m gives
$$\sum_{m=1}^n p_{n,m}^2 \leq \max_k p_{n,k} \left(\sum_m p_{n,m} \right)$$

The first term on the right goes to 0 by assumption. The second is $\lambda_n \to \lambda$. Since we have assumed $\lambda < \infty$, the product of the two terms converges to $0 \cdot \lambda = 0$. □

Intuitively (2.6) says that if we have n people, with probability $p_{n,m}$ to be involved in an accident on a given day, and all of the individual probabilities are small, then the number of accidents that occur has approximately a Poisson distribution with mean $\lambda_n = \sum_m p_{n,m}$. Thus in our Wendy's example we do not need to make the strong assumption that each person has the exact same probability, but only the more reasonable one that the maximum probability is small. Of course if there is one person who always goes to Wendy's on Friday's after the end of his 11:15–12:05 class, then the number of arrivals will not be Poisson.

The last paragraph handles the first problem we have mentioned. To address the problem of varying arrival rates, we generalize the definition.

(2.7) Nonhomogeneous Poisson processes. *We say that $\{N(s), s \geq 0\}$ is a Poisson process with rate $\lambda(r)$ if*

(i) $N(0) = 0$,

(ii) $N(t)$ has independent increments, and

(iii) $N(t+s) - N(s)$ is Poisson with mean $\int_s^t \lambda(r)\, dr$.

The first definition does not work well in this setting since the interarrival times $t_1, t_2, \ldots$ are no longer exponentially distributed or independent. To demonstrate the first claim, we note that

(2.8) $$P(t_1 > t) = P(N(t) = 0) = e^{-\int_0^t \lambda(s)\, ds}$$

since the last expression gives the probability a Poisson with mean $\mu(t) = \int_0^t \lambda(s)\, ds$ is equal to 0. Differentiating gives the density function

$$P(t_1 = t) = -\frac{d}{dt} P(t_1 > t) = \lambda(t) e^{-\int_0^t \lambda(s)\, ds} = \lambda(t) e^{-\mu(t)}$$

Generalizing the last computation shows that the joint distribution

(2.9) $$P(T_1 = u, T_2 = v) = \lambda(u)e^{-\mu(u)} \cdot \lambda(v)e^{-(\mu(v)-\mu(u))}$$

Changing variables,

$$P(t_1 = s, t_2 = t) = \lambda(s)e^{-\mu(s)} \cdot \lambda(s+t)e^{-(\mu(s+t)-\mu(s))}$$

so t_1 and t_2 are not independent when $\lambda(s)$ is not constant. (2.9) generalizes easily to give

$$P(T_1 = v_1, \ldots T_n = v_n) = \lambda(v_1) \cdots \lambda(v_n) e^{-\mu(v_n)}$$

In some applications it is important to have an arrival rate that depends on time. At the Wendy's we have been considering the arrival rate should be fairly small at 11:00, increase to a peak at about 12:30, and go back to a low level by 2:00. However, working with nonhomogeneous processes greatly complicates calculations, so we will usually restrict our attention to the temporally homogeneous case $\lambda(s) \equiv \lambda$. For a concrete example of a nonhomogeneous Poisson process consider the following:

EXERCISE 2.2. Harry's boutique is a trendy clothing store on Rodeo Drive in Beverly Hills, California. Suppose the arrival rate starts at 0 at 10:00, increases to 4 at 12:00, to 6 by 2:00, drops to 2 by 4:00 and decreases to 0 by the time the store closes at 6:00, and that the arrival rates are linear in between these time points. (a) What is the distribution of the number of arrivals in a day? (b) What is the probability no one arrives before noon?

EXERCISE 2.3. Suppose that Harry closes the shop at 5:30 instead of 6:00. (a) What is the expected number of customers that are lost? (b) What is the probability that at least one customer arrives to find the store closed.

3.3. Compound Poisson Processes

In this section we will embellish our Poisson process by associating an independent and identically distributed (i.i.d.) random variable Y_i with each arrival. By independent we mean that the Y_i are independent of each other and of the Poisson process of arrivals. To explain why we have chosen these assumptions, we begin with two examples for motivation.

Example 3.1. Consider the McDonald's restaurant on Route 13 in the southern part of Ithaca. By arguments in the last section, it is not unreasonable to assume that between 12:00 and 1:00 cars arrive according to a Poisson process with rate λ. Let Y_i be the number of people in the ith vehicle. There might be some correlation between the number of people in the car and the arrival time, e.g., more families come to eat there at night, but for a first approximation it seems reasonable to assume that the Y_i are i.i.d. and independent of the Poisson process of arrival times.

Example 3.2. Messages arrive at a central computer to be transmitted across the Internet. If we imagine a large number of users working at terminals connected to a central computer, then the arrival times of messages can be modeled by a Poisson process. If we let Y_i be the size of the ith message, then again it is reasonable to assume $Y_1, Y_2, \ldots$ are i.i.d. and independent of the Poisson process of arrival times.

Having introduced the Y_i's, it is natural to consider the sum of the Y_i's we have seen up to time t:

$$S(t) = Y_1 + \cdots + Y_{N(t)}$$

where we set $S(t) = 0$ if $N(t) = 0$. In Example 3.1, $S(t)$ gives the number of customers that have arrived up to time t. In Example 3.2, $S(t)$ represents the total number of bytes in all of the messages up to time t. In each case it is interesting to know the mean and variance of $S(t)$.

(3.1) Theorem. Let $Y_1, Y_2, \ldots$ be independent and identically distributed, let N be an independent nonnegative integer valued random variable, and let $S = Y_1 + \cdots + Y_N$ with $S = 0$ when $N = 0$.

(i) If $EN < \infty$, then $ES = EN \cdot EY_i$.

(ii) If $EN^2 < \infty$, then $\text{var}(S) = EN \text{var}(Y_i) + \text{var}(N)(EY_i)^2$.

(iii) If N is Poisson(λ), then $\text{var}(S) = \lambda EY_i^2$.

Why is this reasonable? The first of these is natural since if $N = n$ is nonrandom $ES = nEY_i$. (i) then results by setting $n = EN$. The formula in (ii) is more complicated but it clearly has two of the necessary properties:

If $N = n$ is nonrandom, $\text{var}(S) = n\text{var}(Y_i)$.

If $Y_i = c$ is nonrandom $\text{var}(S) = c^2 \text{var}(N)$.

Combining these two observations, we see that $EN \text{var}(Y_i)$ is the contribution to the variance from the variability of the Y_i, while $\text{var}(N)(EY_i)^2$ is the contribution from the variability of N.

Section 3.3 Compound Poisson Processes

Proof. When $N = n$, $S = X_1 + \cdots + X_n$ has $ES = nEY_i$. Breaking things down according to the value of N,

$$ES = \sum_{n=0}^{\infty} E(S|N=n) \cdot P(N=n)$$

$$= \sum_{n=0}^{\infty} nEY_i \cdot P(N=n) = EN \cdot EY_i$$

For the second formula we note that when $N = n$, $S = X_1 + \cdots + X_n$ has $\text{var}(S) = n \text{var}(Y_i)$ and hence,

$$E(S^2|N=n) = n\text{var}(Y_i) + (nEY_i)^2$$

Computing as before we get

$$ES^2 = \sum_{n=0}^{\infty} E(S^2|N=n) \cdot P(N=n)$$

$$= \sum_{n=0}^{\infty} \{n \cdot \text{var}(Y_i) + n^2(EY_i)^2\} \cdot P(N=n)$$

$$= (EN) \cdot \text{var}(Y_i) + EN^2 \cdot (EY_i)^2$$

To compute the variance now, we observe that

$$\text{var}(S) = ES^2 - (ES)^2$$
$$= (EN) \cdot \text{var}(Y_i) + EN^2 \cdot (EY_i)^2 - (EN \cdot EY_i)^2$$
$$= (EN) \cdot \text{var}(Y_i) + \text{var}(N) \cdot (EY_i)^2$$

where in the last step we have used $\text{var}(N) = EN^2 - (EN)^2$ to combine the second and third terms.

Turning to part (iii), we note that in the special case of the Poisson, we have $EN = \lambda$ and $\text{var}(N) = \lambda$, so

$$\text{var}(S) = \lambda \cdot \left(\text{var}(Y_i) + (EY_i)^2\right) = \lambda EY_i^2 \qquad \square$$

For a concrete example of the use of (3.1) consider

Example 3.3. Suppose that the number of customers at a liquor store in a day has a Poisson distribution with mean 81 and that each customer spends an average of $8 with a standard deviation of $6. It follows from (i) in (3.1) that the mean revenue for the day is $81 \cdot \$8 = \$648$. Using (iii) of (3.1), we see that the variance of the total revenue is

$$81 \cdot \{(\$6)^2 + (\$8)^2\} = 8100$$

Taking square roots we see that the standard deviation of the revenue is \$90 compared with a mean of \$648. □

3.4. Thinning and Superposition

In Section 3.3, we added up the Y_i's associated with the arrivals in our Poisson process to see how many customers, etc., we had accumulated by time t. In this section we will use the Y_i to split our one Poisson process into several. Let $N_j(t)$ be the number of $i \leq N(t)$ with $Y_i = j$. In Example 3.1., where Y_i is the number of people in the ith car, $N_j(t)$ will be the number of cars that have arrived by time t with exactly j people. The somewhat remarkable fact is:

(4.1) Theorem. $N_j(t)$ are independent Poisson processes with rate $\lambda P(Y_i = j)$.

Why is this remarkable? There are two "surprises" here: the resulting processes are Poisson and they are independent. To drive the point home consider a Poisson process with rate 10 per hour, and then flip coins to determine whether the arriving customers are male or female. One might think that seeing 40 men arrive in one hour would be indicative of a large volume of business and hence a larger than normal number of women, but (4.1) tells us that the number of men and the number of women that arrive per hour are independent.

Why is this true? For simplicity, we will suppose that $P(Y_i = 1) = p$ and $P(Y_i = 2) = 1 - p$, so there are only two Poisson processes to consider: $N_1(t)$ and $N_2(t)$. We will check the second definition given in (2.4). It should be clear that the independent increments property of the Poisson process implies that the pairs of increments

$$(N_1(t_i) - N_1(t_{i-1}), N_2(t_i) - N_2(t_{i-1})), \quad 1 \leq i \leq n$$

are independent of each other. Since $N_1(0) = N_2(0) = 0$ by definition, it only remains to check that the components $Y = N_1(t+s) - N_1(s)$ and $Z = N_2(t+s) - N_2(s)$ are independent and have the right Poisson distributions. To do this, we note that if $Y = j$ and $Z = k$, then there must have been $j+k$ arrivals between s and $s+t$, j of which were assigned 1's and k of which were assigned 2's, so

$$P(Y = j, Z = k) = e^{-\lambda t} \frac{(\lambda t)^{j+k}}{(j+k)!} \cdot \frac{(j+k)!}{j!k!} p^j (1-p)^k$$

$$= e^{-\lambda p t}\frac{(\lambda p t)^j}{j!}e^{-\lambda(1-p)t}\frac{(\lambda(1-p)t)^k}{k!} \qquad \square$$

Taking one Poisson process and splitting it into two or more by using an i.i.d. sequence Y_i is called **thinning**. Going in the other direction and adding up a lot of independent processes is called **superposition**. Since a Poisson process can be split into independent Poisson processes, it should not be too surprising that when the independent Poisson processes are put together, the sum is Poisson with a rate equal to the sum of the rates.

(4.2) Theorem. *Suppose $N_1(t), \ldots N_k(t)$ are independent Poisson processes with rates $\lambda_1, \ldots, \lambda_k$, then $N_1(t) + \cdots + N_k(t)$ is a Poisson process with rate $\lambda_1 + \cdots + \lambda_k$.*

Why is this true? Again we consider only the case $k = 2$ and check the second definition given in (2.4). It is clear that the sum has independent increments and $N_1(0) + N_2(0) = 0$. To check that the increments have the right Poisson distribution, we note that if $Y = N_1(t+s) - N_1(s)$ and $Z = N_2(t+s) - N_2(s)$, then

$$P(Y + Z = n) = \sum_{m=0}^{n} P(Y = m) P(Z = n - m)$$

$$= \sum_{m=0}^{n} e^{-\lambda_1 t}\frac{(\lambda_1 t)^m}{m!} \cdot e^{-\lambda_2 t}\frac{(\lambda_2 t)^{n-m}}{(n-m)!}$$

Knowing the answer we want, we can rewrite the last expression as

$$e^{-(\lambda_1+\lambda_2)t}\frac{(\lambda_1 t + \lambda_2 t)^n}{n!} \cdot \sum_{m=0}^{n} \binom{n}{m}\left(\frac{\lambda_1}{\lambda_1+\lambda_2}\right)^m \left(\frac{\lambda_2}{\lambda_1+\lambda_2}\right)^{n-m}$$

The sum is 1, since it is the sum of all the probabilities for a binomial(n, p) distribution with $p = \lambda_1/(\lambda_1 + \lambda_2)$. The term outside the sum is the desired Poisson probability, so have proved the desired result. $\square$

We will see in Section 4.1 that the ideas of compounding and thinning are very useful in computer simulations of continuous time Markov chains. For the moment we will illustrate their use in computing the outcome of races between Poisson processes.

Example 3.4. A Poisson race. Given a Poisson process of red arrivals with rate λ and an independent Poisson process of green arrivals with rate μ, what is the probability that we will get 6 red arrivals before a total of 4 green ones?

Solution. The first step is to note that the event in question is equivalent to having at least 6 red arrivals in the first 9. If this happens, then we have at most 3 green arrivals before the 6th red one. On the other hand if there are 5 or fewer red arrivals in the first 9, then we have had at least 4 red arrivals and at most 5 green.

Viewing the red and green Poisson processes as being constructed by starting with one rate $\lambda + \mu$ Poisson process and flipping coins with probability $p = \lambda/(\lambda + \mu)$ to decide the color, we see that the probability of interest is

$$\sum_{k=6}^{9} \binom{9}{k} p^k (1-p)^{9-k}$$

If we suppose for simplicity that $\lambda = \mu$ so $p = 1/2$, this expression becomes

$$\frac{1}{512} \cdot \sum_{k=6}^{9} \binom{9}{k} = \frac{1 + 9 + (9 \cdot 8)/2 + (9 \cdot 8 \cdot 7)/3!}{512} = \frac{140}{512} = 0.273$$

EXERCISE 4.1. *General formula for Poisson races.* Show that the probability that we will get m red arrivals before a total of n green ones is

$$\sum_{k=m}^{m+n-1} \binom{m+n-1}{k} p^k (1-p)^{(m+n-1)-k}$$

3.5. Conditioning

Let $T_1, T_2, T_3, \ldots$ be the arrival times of a Poisson process with rate λ and let $U_1, U_2, \ldots U_n$ be independent and uniformly distributed on $[0, t]$. This section is devoted to the proof of the following remarkable fact.

(5.1) Theorem. *If we condition on $N(t) = n$, then the set of arrival times $\{T_1, \ldots, T_n\}$ has the same distribution as $\{U_1, \ldots, U_n\}$.*

In words, given that there were n arrivals, the set of arrival times is the same as the locations of n darts thrown at random on the interval $[0, t]$.

Why is this true? We begin by finding

$$P(T_1 = v_1, T_2 = v_2, T_3 = v_3 | N(4) = 3)$$

i.e., the joint density function of (T_1, T_2, T_3) given that there were 3 arrivals before time 4. The probability is 0 unless $0 < v_1 < v_2 < v_3 < 4$. To compute the answer in this case, we use the definition of conditional probability $P(B|A) = P(B \cap A)/P(A)$, and rewrite $B \cap A$ in terms of the interarrival times t_1, t_2, t_3, and t_4, to get

$$P(t_1 = v_1, t_2 = v_2 - v_1, t_3 = v_3 - v_2, t_4 > 4 - v_3)/P(N(4) = 3)$$

$$= \frac{\lambda e^{-\lambda v_1} \cdot \lambda e^{-\lambda(v_2 - v_1)} \cdot \lambda e^{-\lambda(v_3 - v_2)} \cdot e^{-\lambda(4 - v_3)}}{e^{-4\lambda} 4^3/3!}$$

$$= \frac{\lambda^3 e^{-4\lambda}}{e^{-4\lambda} 4^3/3!} = \frac{3!}{4^3}$$

Note that the answer does not depend on the values of v_1, v_2, v_3 (as long as $0 < v_1 < v_2 < v_3 < 4$), so the resulting conditional distribution is uniform over

$$\{(v_1, v_2, v_3) : 0 < v_1 < v_2 < v_3 < 4\}$$

This set has volume $4^3/3!$ since $\{(v_1, v_2, v_3) : 0 < v_1, v_2, v_3 < 4\}$ has volume 4^3 and $v_1 < v_2 < v_3$ is one of 3! possible orderings.

Generalizing from the concrete example it is easy to see that if we have times $0 < v_1 < \ldots < v_n < t$, then

(5.2) $$P(T_1 = v_1, \ldots, T_n = v_n | N(t) = n) = n!/t^n$$

while the joint density is 0 otherwise.

To get from (5.2) to (5.1), let $U_1, \ldots, U_n$ be independent and uniform on $(0, t)$ and let $V_1 < V_2 < \ldots < V_n$ be $U_1, U_2, \ldots, U_n$ sorted in increasing order. The V_i are called the **order statistics** for the original sample. The U_i have joint density

$$P(U_1 = u_1, \ldots, U_n = u_n) = \frac{1}{t^n} \quad \text{for } 0 < u_1, \ldots, u_n < t$$

and sorting maps $n!$ vectors to 1, so

$$P(V_1 = v_1, \ldots, V_n = v_n) = \frac{n!}{t^n} \quad \text{for } 0 < v_1 < \ldots < v_n < t$$

The last equation shows that $(T_1, \ldots, T_n)$ and $(V_1, \ldots, V_n)$ have the same distribution. It is clear that the set of values $\{V_1, \ldots, V_n\}$ is the same as $\{U_1, \ldots, U_n\}$, so our proof of (5.1) is complete. □

Theorem (5.1) implies that if we condition on having n arrivals at time t, then the locations of the arrivals are the same as the location of n points thrown uniformly on $[0, t]$. From the last observation we immediately get:

144 Chapter 3 Poisson Processes

(5.3) Theorem. *If $s < t$ and $0 \leq m \leq n$, then*

$$P(N(s) = m | N(t) = n) = \binom{n}{m} \left(\frac{s}{t}\right)^m \left(1 - \frac{s}{t}\right)^{n-m}$$

That is, the conditional distribution of $N(s)$ given $N(t) = n$ is binomial$(n, s/t)$.

Proof from (5.2). The number of arrivals by time s is the same as the number of $U_i < s$. The events $\{U_i < s\}$ these events are independent and have probability s/t, so the number of $U_i < s$ will be binomial$(n, s/t)$. □

Proof by computation. Recalling the definition of conditional probability and rewriting the numerator to take advantage of the fact that the Poisson process has independent increments, we have

$$P(N(s) = m | N(t) = n) = \frac{P(N(s) = m, N(t) - N(s) = n - m)}{P(N(t) = n)}$$

$$= \frac{P(N(s) = m) \cdot P(N(t) - N(s) = n - m)}{P(N(t) = n)}$$

Filling in the probabilities on the right-hand side gives

$$= e^{-\lambda s} \frac{(\lambda s)^m}{m!} \cdot e^{-\lambda(t-s)} \frac{(\lambda(t-s))^{n-m}}{(n-m)!} \Big/ e^{-\lambda t} \frac{(\lambda t)^n}{n!}$$

Performing many cancelations reduces the last mess to

$$= \frac{n!}{m!(n-m)!} \left(\frac{s}{t}\right)^m \left(1 - \frac{s}{t}\right)^{n-m} \qquad \square$$

The rest of this section is devoted to generalizing the results above to a nonhomogeneous Poisson processes with arrival rate $\lambda(r)$. Let $\mu(t) = \int_0^t \lambda(r)\,dr$ and

$$g(r) = \begin{cases} \lambda(r)/\mu(t) & \text{for } 0 < r < t \\ 0 & \text{otherwise} \end{cases}$$

(5.4) Theorem. *Let $U_1, U_2, \ldots U_n$ be independent with density function g. Show that if we condition on $N(t) = n$, then the set of arrival times $\{T_1, \ldots, T_n\}$ has the same distribution as $\{U_1, \ldots, U_n\}$.*

Proof. Again we begin by finding

$$P(T_1 = v_1, T_2 = v_2, T_3 = v_3 | N(4) = 3)$$

Reasoning as above and using Exercise 2.2 we have for $0 < v_1 < v_2 < v_3 < 4$,

$$P(T_1 = v_1, T_2 = v_2, T_3 = v_3, T_4 > 4)$$
$$= \lambda(v_1)e^{-\mu(v_1)} \cdot \lambda(v_2)e^{-(\mu(v_2)-\mu(v_1))} \cdot \lambda(v_3)e^{-(\mu(v_3)-\mu(v_2))} \cdot e^{-(\mu(4)-\mu(v_3))}$$

Simplifying the numerator by combining the exponentials then dividing by $P(N(4) = 3) = e^{-\mu(4)}\mu(4)^3/3!$, the above becomes

$$= \frac{\lambda(v_1)\lambda(v_2)\lambda(v_3)e^{-\mu(4)}}{e^{-\mu(4)}\mu(4)^3/3!} = 3!g(v_1)g(v_2)g(v_3)$$

If we let $V_1 < V_2 < V_3$ be the U_1, U_2, U_3 sorted in increasing order, then (V_1, V_2, V_3) has joint density $3!g(v_1)g(v_2)g(v_3)$. This proves the result when $n = 3$. The general case is similar, only messier to write down. □

We leave the generalization of (5.3) to the reader.

EXERCISE 5.1. Let $N(s)$ be a nonhomogeneous Poisson processes with arrival rate $\lambda(r)$, and let $\mu(t) = \int_0^t \lambda(r)\,dr$. Imitate the "proof by computation" of (5.3) to show that if $s < t$ and $0 \le m \le n$, then

$$P(N(s) = m | N(t) = n) = \binom{n}{m}\left(\frac{\mu(s)}{\mu(t)}\right)^m \left(1 - \frac{\mu(s)}{\mu(t)}\right)^{n-m}$$

3.6. Spatial Poisson Processes

In this section we will construct Poisson processes in the plane and illustrate why they are useful. We begin by revisiting the idea of thinning a Poisson process discussed in Section 3.4. Let $T_1, T_2, \ldots$ be the arrival times of a Poisson processes with rate λ. Let $Y_1, Y_2, \ldots$ be an independent i.i.d. sequence of integer valued random variables, and mark the points (T_i, Y_i) in the plane. (4.1) implies that the points with $Y_i = j$, that is, the points on the line $\{(x, y) : y = j\}$ are a Poisson process, and the Poisson processes for different lines are independent.

Example 6.1. Poisson process in a strip. If we replace the integer valued Y_i by random variables U_i that are uniform on $(0, 1)$, then the preceding construction produces a sequence of points in the strip $\{(t, u) : 0 \le t < \infty, 0 \le u \le 1\}$.

146 Chapter 3 Poisson Processes

In this case since the U_i have a continuous distribution, the number of points on a fixed line $\{(t, u) : u = x\}$ will be 0 with probability one, so we look at things in a different way. Our first step is to fatten the lines up to strips:

$$\{(t, u) : 0 \leq t < \infty, a_{m-1} < u \leq a_m\}$$

and note (4.1) implies that if $0 = a_0 < a_1 \ldots < a_n = 1$, then the points in the strips $1 \leq m \leq n$ are independent.

The last result about disjoint strips when combined with the independent increments property of the individual independent Poisson processes leads to a result about rectangles.

(6.1) Theorem. Let $R_m = \{(t, u) : a_m < t \leq b_m, c_m < u \leq d_m\}$ be rectangles with $a_m \geq 0$ and $0 \leq c_m < d_m \leq 1$, and let N_m be the number of points (T_i, Y_i) that lie in R_m. If the rectangles R_m do not overlap, then the N_m are independent and have a Poisson distribution with mean

$$\mu_m = \lambda(b_m - a_m)(d_m - c_m)$$

That is, λ times the area of R_m.

Why is (6.1) true? Consider first the case of two rectangles. The simplest situation is when the rectangles lie in different strips. If $c_1 < d_1 \leq c_2 < d_2$, so by our remark about disjoint strips, N_1 and N_2 are independent. At the other extreme is the case $c_1 = c_2$ and $d_1 = d_2$, so that the rectangles lie in the same strip. In this situation the independent increments property for the Poisson arrivals with $Y_i \in (c_1, d_1]$ implies that N_1 and N_2 are independent. This leaves us only the awkward case in which the two intervals overlap but are not equal. There are two cases to consider:

Section 3.6 Spatial Poisson Processes   **147**

In either situation we can divide the two rectangles into a total of 4 so that each pair either lies in the same strip or in disjoint strips. Since the number of points in each of the four subrectangles is independent Poisson by the earlier arguments, the result follows from adding up the subrectangles inside each of the rectangles.

When the number of sets gets large then the dissection process for the rectangles gets complicated but the idea behind the proof stays the same. By making enough cuts we can divide the rectangles into subrectangles so that their sides in the y direction are either the same or disjoint. Rather than describe this algorithmically we invite the reader to do it for the set of four rectangles drawn above. □

Technical detail. Note that our rectangles

$$R_m = \{(t,u) : a_m < t \le b_m,\ c_m < u \le d_m\}$$

are closed on one side and open on the other. Thus two rectangles can touch without overlapping and not leave out any points in between. This choice is useful for the proof of (6.3) and is consistent with the usual convention for distribution functions. In many cases the probability of a point on the boundary of the rectangle will be zero, so we don't have to worry about whether to include the sides or not. □

The property in (6.1) is the key to defining the (spatially homogeneous) **Poisson process in the plane**. Given a rectangle A, we let $N(A)$ be the number of points in A. We say that N is a Poisson process with intensity λ if whenever

$$R_m = \{(t,u) : a_m < t \le b_m,\ c_m < u \le d_m\}$$

are nonoverlapping rectangles, then the counts $N(R_m)$ are independent and have mean $\lambda(b_m - a_m)(d_m - c_m)$. Whenever one makes a definition of a new process there is always the question:

Does such a process exist? The answer, of course, is yes. The construction that led us to (6.1) allows us to construct the Poisson process in the unit cube $(0,1] \times (0,1]$. Now divide the plane into a grid of cubes $(i, i+1] \times (j, j+1]$ where i and j are integers, and put independent Poisson processes in each one. By intersecting the rectangle R_m with the cubes in the grid it is easy to see that the number of points in disjoint rectangles are independent Poissons with the correct means. □

One of the most famous data sets fitting the two-dimensional Poisson process is:

Example 6.2. Flying bomb hits in London in World War II. An area in London was divided into $N = 576 = 24 \cdot 24$ areas of 1/4 square kilometers each. This area suffered 535 flying bomb hits, an average of $535/576 = 0.9288$ per square. The next table gives the number of squares, N_k, that were hit exactly k times and compares with the expected number if the hits in each square have a Poisson distribution with this mean.

k	0	1	2	3	4	≥ 5
N_k	229	211	93	35	7	1
Poisson	226.74	211.39	98.54	30.62	7.14	1.57

The fit not only looks good, but to quote Feller, "as judged by the χ^2 criterion, under ideal conditions some 88 per cent of comparable observation should show worse agreement." The quote and this example are from Feller (1968), page 161.

For a more peaceful example, we will now consider traffic on a road with a lot of lanes and not very much traffic. To set up this model we have to first consider:

Example 6.3. Poisson process on the line. Let $T_1, T_2, \ldots$ and $T'_1, T'_2, \ldots$ be independent rate lambda Poisson processes and then mark points at T_i and $-T'_i$. Given an interval $I = (a, b]$, let $N(I)$ be the number of points in I. We leave it to the reader to check that for the resulting process, if $I_m = (a_m, b_m]$ are nonoverlapping intervals, then the counts $N(I_m)$ are independent. For some applications, like the next one, it is useful to renumber the points to use all the integers as indices. That is, we renumber the points so that

$$\ldots T_{-2} < T_{-1} < 0 \leq T_0 < T_1 < T_2 \ldots$$

Example 6.4. Traffic on a wide road. The road has to be wide since we assume that each car has its own preferred velocity that stays constant independent of the other traffic near it on the road. Suppose that initially there is a car at each point T_i, $-\infty < i < \infty$ of a Poisson process on the line defined in the previous example. The points T_i mark the initial locations of the cars on our infinite road. To make them move, we will assign i.i.d. velocities V_i, $-\infty < i < \infty$. Suppose first that the velocities V_i take positive integer values that represent the speed to the right of each car on the road measured in miles per hour. In this case the cars with velocity v are a Poisson process with rate $\lambda P(V_i = v)$.

Obviously a car with velocity v will be in $(a, b]$ at time t if and only if it was in $(a - vt, b - vt]$ at time 0. Since the cars with velocity v all move in parallel it is clear that their initial Poisson distribution will always stay Poisson and hence the number of points in $(a, b]$ at time t with velocity v will be Poisson with mean $\lambda(b-a)P(V_i = v)$. Summing over all v, and recalling that cars with different velocities are independent, it follows that the total number of cars in $(a, b]$ at time t is Poisson with mean $\lambda(b-a)$.

A little more thought reveals that since the Poisson processes of cars of different velocities are independent, then the velocities of the cars in the interval (a, b) are independent of their location and have the same distribution as V_i. Thus

(6.2) Theorem. *If we start cars on the line at points of a Poisson process with rate λ and assign them positive integer-valued independent velocities, then at any time t the positions of the cars are distributed like a rate λ Poisson process and the velocities of the car are independent of their location.*

In words, Poisson locations with independent velocities is an equilibrium state for the traffic model.

As the reader might guess, the last result is true if we assume instead that the velocities V_i have a continuous distribution with density function $g(v)$. To prove this we need to generalize our definition of Poisson process in the plane.

Example 6.5. Nonhomongeneous Poisson process in the plane. Given a rectangle A, let $N(A)$ be the number of points in A. We say that N is a Poisson process with intensity $\lambda(x, y)$ if whenever $R_m = \{(t, u) : a_m < x \le b_m, c_m < y \le d_m\}$ are nonoverlapping rectangles, then the counts $N(R_m)$ are independent and have mean

$$\int_{x=a}^{b} \int_{y=c}^{d} \lambda(x, y) \, dy \, dx$$

In the case of cars on the road $\lambda(x, y) = \lambda g(y)$, where $g(y)$ is the density function for the velocity distribution. To compute the number of cars in (a, b) at time t we need the following:

(6.3) Theorem. *Let A be an open set. The number of points in A is Poisson with mean*
$$\iint_{(x,y)\in A} \lambda(x,y)\, dy\, dx$$
If the integral is ∞, then the number of points is ∞.

Why is this true? If A is a finite disjoint union of rectangles $R_1, \ldots R_m$, this is true since the number in each rectangle are independent Poisson. With a little patience one can write any open set as a union of nonoverlapping squares $R_1, R_2, \ldots$ (see Exercise 6.1). The result then follows by noting that the number of points in $\cup_{m=1}^n R_m$ is Poisson and then letting $n \to \infty$. □

EXERCISE 6.1. Let G be an open set. For each x let $Q(x)$ be the largest square of the form $(i/2^n, (i+1)/2^n] \times (j/2^n, (j+1)/2^n]$ that contains x. Show that (a) $\cup_{x\in G} Q(x) = G$. (b) If $y \in Q(x)$ then $Q(y) = Q(x)$. (c) Conclude that $\{Q(x) : x \in G\}$ gives the desired covering.

Application to Example 6.4. The points in (a,b) at time t are those whose initial position x and velocity v satisfy $x + vt \in (a,b)$ or $x \in (a - vt, b - vt)$. Using (6.3) now shows that the number is Poisson with mean
$$\int_{v=0}^{\infty} \int_{x=a-vt}^{b-vt} \lambda g(v)\, dx\, dv = \int_{v=0}^{\infty} \lambda(b-a) g(v)\, dv = \lambda(b-a)$$

where the first equality follows from the fact that the integrand does not depend on x; the second, from the fact that $g(v)$ is a density and hence integrates to 1.

As another application of (6.3) we will now consider

Example 6.6. Nonhomogeneous thinning. Let $T_1, T_2, \ldots$ be a Poisson process with rate λ.

(6.4) Theorem. *Suppose that in a Poisson process with rate λ, we keep a point that lands at s with probability $p(s)$. Then the result is a nonhomogeneous Poisson process with rate $\lambda p(s)$.*

Proof. Associate with each arrival T_i, an independent random variable U_i that is uniform on $(0,1)$, and accept the point if $U_i < p(T_i)$. The number of points T_i accepted in an interval (a,b) is the same as the number of (T_i, U_i) that land in the region
$$\{(t,u) : a < t < b, 0 < u < p(t)\}$$

so by (6.3) the number is Poisson with mean λ times the area of this set, which is $\lambda \int_{t=a}^{b} p(t)\, dt$. Clearly the number of arrivals in disjoint intervals are independent, so we have checked the definition of nonhomogeneous Poisson process given in (2.7). □

For an application of (6.4) consider

Example 6.7. M/G/∞ queue. In modeling telephone traffic, we can, as a first approximation, suppose that the number of phone lines is infinite, i.e., everyone who tries to make a call finds a free line. This certainly is not always true but developing a model in which we pretend this is true can help us to discover how many phone lines we need to be able to provide service 99.99% of the time.

The argument for arrivals at Wendy's implies that the beginnings of calls follow a Poisson process. As for the calls themselves there is no reason to suppose that their duration follows a special distribution like the exponential, so use a general distribution function G with $G(0) = 0$ and mean μ. Suppose that the system starts empty at time 0. The probability a call started at s has ended by time t is $G(t - s)$, so using (6.4) the number of calls still in progress at time t is Poisson with mean

$$\int_{s=0}^{t} \lambda(1 - G(t-s))\, ds = \lambda \int_{r=0}^{r} (1 - G(r))\, dr$$

Letting $t \to \infty$ we see that in the long run the number of calls in the system will be Poisson with mean

$$\lambda \int_{r=0}^{\infty} (1 - G(r))\, dr = \lambda \mu$$

That is, the mean number in the system is the rate at which calls enter times their average duration.

In the argument above we supposed that the system starts empty. Since the number of initial calls in the system at time t decreases to 0 as $t \to \infty$, the limiting result is true for any initial number of calls X_0. □

The thinning trick in Example 6.7 can also be used for:

Example 6.8. Constructing nonhomogeneous Poisson processes. Suppose $\lambda(t, x) \leq \lambda$. Let $N(t)$ be a homogeneous Poisson process with intensity λ and let $p(t, x) = \lambda(t, x)/\lambda$. Associate with each point (T_i, X_i) an independent random variable U_i uniform on $(0, 1)$, and accept the point if $p(T_i, X_i) < U_i$. It follows from the proof of (6.4) that the process $\hat{N}(t)$ of accepted points is a Poisson process with intensity $\lambda(t, x)$. □

3.7. Exercises

7.1. Suppose that the time to repair a machine is exponentially distributed random variable with mean 2. (a) What is the probability the repair takes more than 2 hours. (b) What is the probability that the repair takes more than 5 hours given that it takes more than 3 hours.

7.2. The lifetime of a radio is exponentially distributed with mean 5 years. If Ted buys a 7 year-old radio, what is the probability it will be working 3 years later?

7.3. Alice and Betty enter a beauty parlor simultaneously, Alice to get a manicure and Betty to get a haircut. Suppose the time for a manicure (haircut) is exponentially distributed with mean 20 (30) minutes. (a) What is the probability Alice gets done first? (b) What is the expected amount of time until Alice and Betty are both done?

7.4. Let S and T be exponentially distributed with rates λ and μ. Let $U = \min\{S, T\}$ and $V = \max\{S, T\}$. Find (a) EU. (b) $E(V - U)$, (c) EV. (d) Use the identity $V = S + T - U$ to get a different looking formula for EV.

7.5. Let T_1 and T_2 be exponentially distributed with rates λ_1 and λ_2. Let $U = \min\{T_1, T_2\}$, $V = \max\{T_1, T_2\}$, and I be the index of the smaller time, i.e., $T_I = U$. Compute the joint density $P(U = x, V - U = y, I = i)$ and use this to conclude that U and $V - U$ are independent.

7.6. Consider a bank with two tellers. Three people, Alice, Betty, and Carol enter the bank at almost the same time and in that order. Alice and Betty go directly into service while Carol waits for the first available teller. Suppose that the service times for each customer are exponentially distributed with mean 4 minutes. (a) What is the expected total amount of time for Carol to complete her businesses? (b) What is the expected total time until the last of the three customers leaves? (c) What is the probability Carol is the last one to leave? (d) Answer questions (a),(b), and (c) for a general exponential with rate λ.

7.7. Consider the set-up of the previous problem but now suppose that the two tellers have exponential service times with means 3 and 6 minutes. Answer questions (a), (b), and (c).

7.8. Consider the set-up of the previous problem but now suppose that the two tellers have exponential service times with rates $\lambda \leq \mu$. Again, answer questions (a), (b), and (c).

7.9. A machine has two critically important parts and is subject to three different types of shocks. Shocks of type i occur at times of a Poisson process with rate λ_i. Shocks of types 1 break part 1, those of type 2 break part 2, while those of type 3 break both parts. Let U and V be the failure times of the two parts. (a) Find $P(U > s, V > t)$. (b) Find the distribution of U and the distribution of V. (c) Are U and V independent?

7.10. Three people are fishing and each catches fish at rate 2 per hour. How long do we have to wait until everyone has caught at least one fish?

7.11. A submarine has three navigational devices but can remain at sea if at least two are working. Suppose that the failure times are exponential with means 1 year, 1.5 years, and 3 years. What is the average length of time the boat can remain at sea.

7.12. Three students arrive at the beginning of a professor's office hours. The amount of time they will stay is exponentially distributed with rates 1, 2, and 3, that is, means of 1, 1/2, and 1/3 hour. (a) What is the expected time until only one student remains? (b) Number the students according to their exponential rates. Find the distribution of $L =$ the departure rate for last student left. (c) What is the expected time until all three students are gone?

7.13. In a hardware store you must first go to server 1 to get your goods and then go to a server 2 to pay for them. Suppose that the times for the two activities are exponentially distributed with means 6 minutes and 3 minutes. Compute the average amount of time it take Bob to get his goods and pay if when he comes in there is one customer named Al with server 1 and no one at server 2.

7.14. Consider the set-up of the previous problem but suppose that the times for the two activities are exponentially distributed with rates λ and μ. Compute Bob's average waiting time.

7.15. A flashlight needs two batteries to be operational. You start with four batteries numbered 1 to 4. Whenever a battery fails it is replaced by the lowest-numbered working battery. Suppose that battery life is exponential with mean 100 hours. Let T be the time at which there is one working battery left and N be the number of the one battery that is still good. (a) Find ET. (b) Find the distribution of N. (c) Solve (a) and (b) for a general number of batteries.

7.16. Customers arrive at a shipping office at times of a Poisson process with rate 3 per hour. (a) The office was supposed to open at 8AM but the clerk Oscar overslept and came in at 10AM. What is the probability that no customers came in the two-hour period? (b) What is the distribution of the amount of time Oscar has to wait until his first customer arrives?

7.17. Suppose $N(t)$ is a Poisson process with rate 2. Compute the conditional probabilities (a) $P(N(3) = 4|N(1) = 1)$, (b) $P(N(1) = 1|N(3) = 4)$.

7.18. Suppose $N(t)$ is a Poisson process with rate 3. Let T_n denote the time of the nth arrival. Find (a) $E(T_{12})$, (b) $E(T_{12}|N(2)=5)$, (c) $E(N(5)|N(2)=5)$.

7.19. Customers arrive at a bank according to a Poisson process with rate 10 per hour. Given that two customers arrived in the first 5 minutes, what is the probability that (a) both arrived in the first 2 minutes. (b) at least one arrived in the first 2 minutes.

7.20. Traffic on Rosedale Road in Princeton, NJ, follows a Poisson process with rate 6 cars per minute. A deer runs out of the woods and tries to cross the road. If there is a car passing in the next 5 seconds then there will be a collision. (a) Find the probability of a collision. (b) What is the chance of a collision if the deer only needs 2 seconds to cross the road. .

7.21. Traffic on Snyder Hill Road in Ithaca, NY, follows a Poisson process with rate 2/3's of a vehicle per minute. 10% of the vehicles are trucks, the other 90% are cars. (a) What is the probability at least one truck passes in a hour? (b) Given that 10 trucks have passed by in an hour, what is the expected number of vehicles that have passed by. (c) Given that 50 vehicles have passed by in a hour, what is the probability there were exactly 5 trucks and 45 cars.

7.22. Suppose that the number of calls per hour to an answering service follows a Poisson process with rate 4. (a) What is the probability that fewer (i.e., <) than 2 calls came in the first hour? (b) Suppose that 6 calls arrive in the first hour, what is the probability there will be < 2 in the second hour. (c) Given that 6 calls arrive in the first two hours, what is the conditional probability exactly 2 arrived in the first hour and exactly 4 in the second? (d) Suppose that the operator gets to take a break after she has answered 10 calls. How long are her average work periods?

7.23. Continuing with the set-up of the previous problem, suppose that 3/4's of the calls are made by men, 1/4 by women, and the sex of the caller is independent of the time of the call. (a) What is the probability that in one hour exactly 2 men and 3 women will call the answering service? (b) What is the probability 3 men will make phone calls before 3 women do?

7.24. Hockey teams 1 and 2 score goals at times of Poisson processes with rates 1 and 2. Suppose that $N_1(0)=3$ and $N_2(0)=1$. (a) What is the probability that $N_1(t)$ will reach 5 before $N_2(t)$ does? (b) Answer part (a) for Poisson processes with rates λ_1 and λ_2.

7.25. Consider two independent Poisson processes $N_1(t)$ and $N_2(t)$ with rates λ_1 and λ_2. What is the probability that the two-dimensional process $(N_1(t), N_2(t))$ ever visits the point (i,j)?

7.26. Rock concert tickets are sold at a ticket counter. Females and males arrive at times of independent Poisson processes with rates 30 and 20. (a) What is the probability the first three customers are female? (b) If exactly 2 customers arrived in the first five minutes, what is the probability both arrived in the first three minutes. (c) Suppose that customers regardless of sex buy 1 ticket with probability 1/2, two tickets with probability 2/5, and three tickets with probability 1/10. Let N_i be the number of customers that buy i tickets in the first hour. Find the joint distribution of (N_1, N_2, N_3).

7.27. Ellen catches fish at times of a Poisson process with rate 2 per hour. 40% of the fish are salmon, while 60% of the fish are trout. What is the probability she will catch exactly 1 salmon and 2 trout if she fishes for 2.5 hours?

7.28. Edwin catches trout at times of a Poisson process with rate 3 per hour. Suppose that the trout weigh an average of 4 pounds with a standard deviation of 2 pounds. Find the mean and standard deviation of the total weight of fish he catches in two hours.

7.29. An insurance company pays out claims at times of a Poisson process with rate 4 per week. Writing K as shorthand for "thousands of dollars," suppose that the mean payment is 10K and the standard deviation is 6K. Find the mean and standard deviation of the total payments for 4 weeks.

7.30. Customers arrive at an automated teller machine at the times of a Poisson process with rate of 10 per hour. Suppose that the amount of money withdrawn on each transaction has a mean of $30 and a standard deviation of $20. Find the mean and standard deviation of the total withdrawals in 8 hours.

7.31. Messages arrive to be transmitted across the internet at times of a Poisson process with rate λ. Let Y_i be the size of the ith message, measured in bytes, and let $g(z) = Ez^{Y_i}$ be the generating function of Y_i. Let $N(t)$ be the number of arrivals at time t and $S = Y_1 + \cdots + Y_{N(t)}$ be the total size of the messages up to time t. (a) Find the generating function $f(z) = E(z^S)$. (b) Differentiate and set $z = 1$ to find ES. (c) Differentiate again and set $z = 1$ to find $E\{S(S-1)\}$. (d) Compute var(S).

7.32. Let $t_1, t_2, \ldots$ be independent exponential(λ) random variables and let N be an independent random variable with $P(N = n) = (1-p)^{n-1}$. What is the distribution of the random sum $T = t_1 + \cdots + t_N$?

7.33. Signals are transmitted according to a Poisson process with rate λ. Each signal is successfully transmitted with probability p and lost with probability $1 - p$. The fates of different signals are independent. For $t \geq 0$ let $N_1(t)$ be the number of signals successfully transmitted and let $N_2(t)$ be the number that are lost up to time t. (a) Find the distribution of $(N_1(t), N_2(t))$. (b) What is the distribution of $L = $ the number of signals lost before the first one is successfully transmitted?

7.34. A copy editor reads a 200-page manuscript, finding 108 typos. Suppose that the author's typos follow a Poisson process with some unknown rate λ per page, while from long experience we know that the copyeditor finds 90% of the mistakes that are there. (a) Compute the expected number of typos found as a function of the arrival rate λ. (b) Use the answer to (a) to find an estimate of λ and of the number of undiscovered typos.

7.35. Two copy editors read a 300-page manuscript. The first found 100 typos, the second found 120, and their lists contain 80 errors in common. Suppose that the author's typos follow a Poisson process with some unknown rate λ per page, while the two copy editors catch errors with unknown probabilities of success p_1 and p_2. Let X_0 be the number of typos that neither found. Let X_1 and X_2 be the number of typos found only by 1 or only by 2, and let X_3 be the number of typos found by both. (a) Find the joint distribution of (X_0, X_1, X_2, X_3). (b) Use the answer to (a) to find an estimates of p_1, p_2 and then of the number of undiscovered typos.

7.36. A light bulb has a lifetime that is exponential with a mean of 200 days. When it burns out a janitor replaces it immediately. In addition there is a handyman who comes at times of a Poisson process at rate .01 and replaces the bulb as "preventive maintenance." (a) How often is the bulb replaced? (b) In the long run what fraction of the replacements are due to failure?

7.37. Calls to the Dryden fire department arrive according to a Poisson process with rate 0.5 per hour. Suppose that the time required to respond to a call, return to the station, and get ready to respond to the next call is uniformly distributed between 1/2 and 1 hour. If a new call comes before the Dryden fire department is ready to respond, the Ithaca fire department is asked to respond. Suppose that the Dryden fire department is ready to respond now. (a) Find the probability distribution for the number of calls they will handle before they have to ask for help from the Ithaca fire department. (b) In the long run what fraction of calls are handled by the Ithaca fire department?

7.38. A math professor waits at the bus stop at the Mittag-Leffler Institute in the suburbs of Stockholm, Sweden. Since he has forgotten to find out about the bus schedule, his waiting time until the next bus is uniform on $(0,1)$. Cars drive by the bus stop at rate 6 per hour. Each will take him into town with probability 1/3. What is the probability he will end up riding the bus?

7.39. Consider a Poisson process with rate λ and let L be the time of the last arrival in the interval $[0,t]$, with $L=0$ if there was no arrival. (a) Compute $E(t-L)$ (b) What happens when we let $t \to \infty$ in the answer to (a)?

7.40. Let T be exponentially distributed with rate λ. (a) Use the definition of conditional expectation to compute $E(T|T<c)$. (b) Determine $E(T|T<c)$

from the identity

$$ET = P(T < c)E(T|T < c) + P(T > c)E(T|T > c)$$

7.41. *When did the chicken cross the road?* Suppose that traffic on a road follows a Poisson process with rate λ cars per minute. A chicken needs a gap of length at least c minutes in the traffic to cross the road. To compute the time the chicken will have to wait to cross the road, let $t_1, t_2, t_3, \ldots$ be the interarrival times for the cars and let $J = \min\{j : t_j > c\}$. If $T_n = t_1 + \cdots + t_n$, then the chicken will start to cross the road at time T_{J-1} and complete his journey at time $T_{J-1} + c$. Use the previous exercise to show $E(T_{J-1} + c) = (e^{\lambda c} - 1)/\lambda$.

7.42. *Chicken à la martingale.* Solve the previous problem by noting that $T_n - (n/\lambda)$ is a martingale and stopping at time J. (No, this is not a typo! $J - 1$ is not a stopping time.)

7.43. Starting at some fixed time, which we will call 0 for convenience, satellites are launched at times of a Poisson process with rate λ. After an independent amount of time having distribution function F and mean μ, the satellite stops working. Let $X(t)$ be the number of working satellites at time t. (a) Find the distribution of $X(t)$. (b) Let $t \to \infty$ in (a) to show that the limiting distribution is Poisson($\lambda\mu$).

7.44. Ignoring the fact that the bar exam is only given twice a year, let us suppose that new lawyers arrive in Los Angeles according to a Poisson process with mean 300 per year. Suppose that each lawyer independently practices for an amount of time T with a distribution function $F(t) = P(T \leq t)$ that has $F(0) = 0$ and mean 25 years. Show that in the long run the number of lawyers in Los Angeles is Poisson with mean 7500.

4 Continuous-Time Markov Chain

4.1. Definitions and Examples

In Chapter 1 we considered Markov chains X_n with a discrete time index $n = 0, 1, 2, \ldots$ In this chapter we will extend the notion to a continuous time parameter $t \geq 0$, a setting that is more convenient for some applications. In discrete time we formulated the Markov property as: for any possible values of $j, i, i_{n-1}, \ldots i_0$

$$P(X_{n+1} = j | X_n = i, X_{n-1} = i_{n-1}, \ldots, X_0 = i_0) = P(X_{n+1} = j | X_n = i)$$

In continuous time, it is technically difficult to define the conditional probability given all of the X_r for $r \leq s$, so we instead say that X_t, $t \geq 0$ is a Markov chain if for any $0 \leq s_0 < s_1 \cdots < s_n < s$ and possible states $i_0, \ldots, i_n, i, j$ we have

$$P(X_{t+s} = j | X_s = i, X_{s_n} = i_n, \ldots, X_{s_0} = i_0) = P(X_t = j | X_0 = i)$$

In words, given the present state, the rest of the past is irrelevant for predicting the future. Note that built into the definition is the fact that the probability of going from i at time s to j at time $s + t$ only depends on t the difference in the times.

Our first step is to construct a large collection of examples. In Example 1.6 we will see that this is almost the general case.

Example 1.1. Let $N(t)$, $t \geq 0$ be a Poisson process with rate λ and let Y_n be a discrete time Markov chain with transition probability $u(i, j)$. Then $X_t = Y_{N(t)}$ is a continuous-time Markov chain. In words, X_t takes one jump according to $u(i, j)$ at each arrival of $N(t)$.

Why is this true? Intuitively, this follows from the lack of memory property of the exponential distribution. If $X_s = i$, then independent of what has happened in the past, the time to the next jump will be exponentially distributed with rate λ and will go to state j with probability $u(i,j)$. □

Discrete time Markov chains were described by giving their transition probabilities $p(i,j)$ = the probability of jumping from i to j in one step. In continuous time there is no first time $t > 0$ so we introduce for each $t > 0$ a **transition probability**

$$p_t(i,j) = P(X_t = j | X_0 = i)$$

To compute this for Example 1.1, we note that $N(t)$ has a Poisson number of jumps with mean λt, so

$$p_t(i,j) = \sum_{n=0}^{\infty} e^{-\lambda t} \frac{(\lambda t)^n}{n!} u^n(i,j)$$

where $u^n(i,j)$ is the nth power of the transition probability $u(i,j)$.

In continuous time, as in discrete time, the transition probability satisfies

(1.1) Chapman–Kolmogorov equation.

$$\sum_k p_s(i,k) p_t(k,j) = p_{s+t}(i,j)$$

Why is this true? In order for the chain to go from i to j in time $s+t$, it must be in some state k at time s, and the Markov property implies that the two parts of the journey are independent. □

Proof. Breaking things down according to the state at time s, we have

$$P(X_{s+t} = j | X_0 = i) = \sum_k P(X_{s+t} = j, X_s = k | X_0 = i)$$

Using the definition of conditional probability and the Markov property, the above is

$$= \sum_k P(X_{s+t} = j | X_s = k, X_0 = i) P(X_s = k | X_0 = i) = \sum_k p_t(k,j) p_s(i,k) \quad \square$$

(1.1) shows that if we know the transition probability for $t < t_0$ for any $t_0 > 0$, we know it for all t. This observation and a large leap of faith (which we

160 Chapter 4 Continuous-Time Markov Chains

will justify later) suggests that the transition probabilities p_t can be determined from their derivatives at 0:

$$(1.2) \qquad q(i,j) = \lim_{h \to 0} \frac{p_h(i,j)}{h} \qquad \text{for } j \neq i$$

If this limit exists (and it will in all the cases we consider) we will call $q(i,j)$ the **jump rate** from i to j. To explain this name we will compute the:

Jump rates for Example 1.1. The probability of at least two jumps by time h is 1 minus the probability of 0 or 1 jumps

$$1 - \left(e^{-\lambda h} + \lambda h e^{-\lambda h}\right) = 1 - (1 + \lambda h)\left(1 - \lambda h + \frac{(\lambda h)^2}{2!} + \ldots\right)$$
$$= (\lambda h)^2/2! + \ldots = o(h)$$

That is, when we divide it by h it tends to 0 as $h \to 0$. The probability of going from i to j in zero steps, $u^0(i,j) = 0$, when $j \neq i$, so

$$\frac{p_h(i,j)}{h} \approx \lambda e^{-\lambda h} u(i,j) \to \lambda u(i,j)$$

as $h \to 0$. Comparing the last equation with the definition of the jump rate in (1.2) we see that $q(i,j) = \lambda u(i,j)$. In words we leave i at rate λ and go to j with probability $u(i,j)$. □

Example 1.1 is atypical. There we started with the Markov chain and then defined its rates. In most cases it is much simpler to describe the system by writing down its transition rates $q(i,j)$ for $i \neq j$, which describe the rates at which jumps are made from i to j. The simplest possible example is:

Example 1.2. Poisson process. Let $X(t)$ be the number of arrivals up to time t in a Poisson process with rate λ. Since arrivals occur at rate λ in the Poisson process the number of arrivals, $X(t)$, increases from n to $n+1$ at rate λ, or in symbols

$$q(n, n+1) = \lambda \quad \text{for all } n \geq 0$$

This simplest example is a building block for other examples:

Example 1.3. M/M/s queue. Imagine a bank with s tellers that serve customers who queue in a single line if all of the servers are busy. We imagine that customers arrive at times of a Poisson process with rate λ, and that each service time is an independent exponential with rate μ. As in Example 1.2, $q(n, n+1) = \lambda$. To model the departures we let

$$q(n, n-1) = \begin{cases} n\mu & 0 \le n \le s \\ s\mu & n \ge s \end{cases}$$

To explain this, we note that when there are $n \le s$ individuals in the system then they are all being served and departures occur at rate $n\mu$. When $n > s$, all s servers are busy and departures occur at $s\mu$. □

This model is in turn a stepping stone to another, more realistic one:

Example 1.4. M/M/s queue with balking. Again customers arrive at the bank in Example 1.3 at rate λ, but this time they only join the queue with probability a_n if there are n customers in line. Since customers flip coins to determine if they join the queue, this thins the Poisson arrival rate so that

$$q(n, n+1) = \lambda a_n \quad \text{for } n \ge 0$$

Of course, the service rates $q(n, n-1)$ remain as they were previously. □

Having seen several examples, it is natural to ask:

Given the rates, how do you construct the chain?

Let $\lambda_i = \sum_{j \ne i} q(i, j)$ be the rate at which X_t leaves i. If $\lambda_i = \infty$, it will want to leave immediately, so we will always suppose that each state i has $\lambda_i < \infty$. If $\lambda_i = 0$, then X_t will never leave i. So suppose $\lambda_i > 0$ and let

$$r(i, j) = q(i, j)/\lambda_i$$

Here r, short for "routing matrix," is the probability the chain goes to j when it leaves i.

Informal construction. If X_t is in a state i with $\lambda_i = 0$ then X_t stays there forever. If $\lambda_i > 0$, X_t stays at i for an exponentially distributed amount of time with rate λ_i, then goes to state j with probability $r(i, j)$.

Formal construction. Suppose, for simplicity, that $\lambda_i > 0$ for all i. Let Y_n be a Markov chain with transition probability $r(i, j)$. The discrete-time chain Y_n, gives the road map that the continuous-time process will follow. To determine how long the process should stay in each state let $\tau_0, \tau_1, \tau_2, \ldots$ be independent exponentials with rate 1.

At time 0 the process is in state X_0 and should stay there for an amount of time that is exponential with rate $\lambda(X_0)$, so we let the time the process stays in state X_0 be $t_1 = \tau_0/\lambda(X_0)$.

At time t_1 the process jumps to X_1, where it should stay for an exponential amount of time with rate $\lambda(X_1)$, so we let the time the process stays in state X_1 be $t_2 = \tau_1/\lambda(X_1)$.

At time $T_2 = t_1 + t_2$ the process jumps to X_2, where it should stay for an exponential amount of time with rate $\lambda(X_2)$, so we let the time the process stays in state X_2 be $t_3 = \tau_2/\lambda(X_2)$.

Continuing in the obvious way, we can let the amount of time the process stays in X_{n-1} be $t_n = \tau_{n-1}/\lambda(X_{n-1})$, so that the process jumps to X_n at time
$$T_n = t_1 + \cdots + t_n$$

In symbols, if we let $T_0 = 0$, then for $n \geq 0$ we have

(1.3) $$X(t) = Y_n \quad \text{for } T_n \leq t < T_{n+1}$$

Computer simulation. Before we turn to the dark side of the construction above, the reader should observe that it gives rise to a recipe for simulating a Markov chain. Generate independent standard exponentials τ_i, say, by looking at $\tau_i = -\ln U_i$ where U_i are uniform on $(0,1)$. Using another sequence of random numbers, generate the transitions of X_n, then define t_i, T_n, and X_t as above.

The good news about the formal construction above is that if $T_n \to \infty$ as $n \to \infty$, then we have succeeded in defining the process for all time and we are done. This will be the case in almost all the examples we consider. The bad news is that $\lim_{n \to \infty} T_n < \infty$ can happen.

Example 1.5. An exploding Markov chain. Think of a ball that is dropped and returns to half of its previous height on each bounce. Summing $1/2^n$ we conclude that all of its infinitely many bounces will be completed in a finite amount of time. To turn this idea into an example of a badly behaved Markov chain on the state space $S = \{1, 2, \ldots\}$, suppose $q(i, i+1) = 2^i$ for $i \geq 1$ with all the other $q(i,j) = 0$. The chain stays at i for an exponentially distributed amount of time with mean 2^{-i} before it goes to $i+1$. Let T_j be the first time the chain reaches j. By the formula for the rates

$$\sum_{i=1}^{\infty} E_1(T_{i+1} - T_i) = \sum_{i=1}^{\infty} 2^{-i} = 1$$

This implies $T_\infty = \lim_{n\to\infty} T_n$ has $E_1 T_\infty = 1$ and hence $P_1(T_\infty < \infty) = 1$. □

In most models, it is senseless to have the process make an infinite amount of jumps in a finite amount of time so we introduce a "cemetery state" Δ to the state space and complete the definition by letting $T_\infty = \lim_{n\to\infty} T_n$ and stetting
$$X(t) = \Delta \quad \text{for all } t \geq T_\infty$$

The simplest way to rule out explosions (i.e., $P_x(T_\infty < \infty) > 0$) is to consider

Example 1.6. Markov chains with bounded rates. When the maximum transition rate
$$\Lambda = \max_i \lambda_i < \infty$$
we can use a trick to reduce the process to one with constant transition rates $\lambda_i \equiv \Lambda$. Let
$$u(i,j) = q(i,j)/\Lambda \quad \text{for } j \neq i$$
$$u(i,i) = 1 - \sum_{j \neq i} q(i,j)/\Lambda$$
In words, while the chain is at any state i, it attempts to make transitions at rate Λ. On each attempt it jumps from i to j with probability $u(i,j)$ and stays put with the remaining probability $1 - \sum_{j \neq i} u(i,j)$. Since the jump rate is independent of the state, the jump times are a Poisson process and $T_n \to \infty$. □

Here, we have come full circle by writing a general Markov chain with bounded flip rates in the form given in Example 1.1. This observation is often useful in simulation. Since the holding times in each state are exponentially distributed with the same rate Λ, then, as we will see in Section 4, if we are interested in the long-run fraction of time the continuous time chain spends in each state, we can ignore the holding times and simulate the discrete time chain with transition probability u.

There are many interesting examples with unbounded rates. Perhaps the simplest and best known is

Example 1.7. The Yule process. In this simple model of the growth of a population (of bacteria for example), there are no deaths and each particle splits into birth at rate β, so $q(i, i+1) = \beta i$ and the other $q(i,j)$ are 0. If we start with one individual, then the jump to $n+1$ is made at time $T_n = t_1 + \cdots + t_n$, where t_n is exponential with rate βn. $E t_n = 1/\beta i$, so
$$E T_n = (1/\beta) \sum_{m=1}^{n} 1/m \sim (\log n)/\beta$$

as $n \to \infty$. This is by itself not enough to establish that $T_n \to \infty$, but it is not hard to fill in the missing details.

Proof. $\text{var}(T_n) = \sum_{m=1}^{n} 1/m^2\beta^2 \le C = \sum_{m=1}^{\infty} 1/m^2\beta^2$. Chebyshev's inequality implies
$$P(T_n \le ET_n/2) \le 4C/(ET_n)^2 \to 0$$
as $n \to \infty$. Since $n \to T_n$ is increasing, it follows that $T_n \to \infty$. □

The next example shows that linear growth of the transition rates is at the borderline of explosive behavior.

Example 1.8. Pure birth processes with power law rates. Suppose $q(i, i+1) = i^p$ and all the other $q(i,j) = 0$. In this case the jump to $n+1$ is made at time $T_n = t_1 + \cdots + t_n$, where t_n is exponential with rate n^p. $Et_n = 1/n^p$, so if $p > 1$
$$ET_n = \sum_{m=1}^{n} 1/m^p$$
stays bounded as $n \to \infty$. This implies $ET_\infty = \sum_{m=1}^{\infty} 1/m^p$, so $T_\infty < \infty$ with probability one.

4.2. Computing the Transition Probability

In the last section we saw that given jump rates $q(i,j)$ we can construct a Markov chain that has these jump rates. This chain, of course, has a transition probability
$$p_t(i,j) = P(X_t = j | X_0 = i)$$
Our next question is: How do you compute the transition probability p_t from the jump rates q?

Our road to the answer starts by using the Chapman–Kolmogorov equations, (1.1), and then taking the $k = i$ term out of the sum.

(2.1)
$$\begin{aligned} p_{t+h}(i,j) - p_t(i,j) &= \left(\sum_k p_h(i,k)p_t(k,j)\right) - p_t(i,j) \\ &= \left(\sum_{k \ne i} p_h(i,k)p_t(k,j)\right) + [p_h(i,i) - 1]\,p_t(i,j) \end{aligned}$$

Our goal is to divide each side by h and let $h \to 0$ to compute
$$p'_t(i,j) = \lim_{h \to 0} \frac{p_{t+h}(i,j) - p_t(i,j)}{h}$$

Section 4.2 Computing the Transition Probability

By the definition of the jump rates

$$q(i,j) = \lim_{h \to 0} \frac{p_h(i,j)}{h} \quad \text{for } i \neq j$$

so ignoring the detail of interchanging the limit and the sum, we have

$$(\star) \qquad \lim_{h \to 0} \frac{1}{h} \sum_{k \neq i} p_h(i,k) p_t(k,j) = \sum_{k \neq i} q(i,k) p_t(k,j)$$

For the other term we note that $1 - p_h(i,i) = \sum_{k \neq i} p_h(i,k)$, so

$$\lim_{h \to 0} \frac{p_h(i,i) - 1}{h} = - \lim_{h \to 0} \sum_{k \neq i} \frac{p_h(i,k)}{h} = -\sum_{k \neq i} q(i,k) = -\lambda_i$$

and we have

$$(\star\star) \qquad \lim_{h \to 0} \frac{p_h(i,i) - 1}{h} p_t(i,j) = -\lambda_i p_t(i,j)$$

Combining $(\star)$ and $(\star\star)$ with (2.1) and the definition of the derivative we have

$$(2.2) \qquad p'_t(i,j) = \sum_{k \neq i} q(i,k) p_t(k,j) - \lambda_i p_t(i,j)$$

To neaten up the last expression we introduce a new matrix

$$Q(i,j) = \begin{cases} q(i,j) & \text{if } j \neq i \\ -\lambda_i & \text{if } j = i \end{cases}$$

For future computations note that the off-diagonal elements $q(i,j)$ with $i \neq j$ are nonnegative, while the diagonal entry is a negative number chosen to make the row sum equal to 0.

Using matrix notation we can write (2.2) simply as

$$(2.3) \qquad p'_t = Q p_t$$

This is **Kolmogorov's backward equation**. If Q were a number instead of a matrix, the last equation would be easy to solve. We would set $p_t = e^{Qt}$ and check by differentiating that the equation held. Inspired by this observation, we define the matrix

$$(2.4) \qquad e^{Qt} = \sum_{n=0}^{\infty} \frac{(Qt)^n}{n!} = \sum_{n=0}^{\infty} Q^n \cdot \frac{t^n}{n!}$$

and check by differentiating that

$$\frac{d}{dt}e^{Qt} = \sum_{n=1}^{\infty} Q^n \frac{t^{n-1}}{(n-1)!} = \sum_{n=1}^{\infty} Q \cdot \frac{Q^{n-1}t^{n-1}}{(n-1)!} = Qe^{Qt}$$

Fine print. Here we have interchanged the operations of summation and differentiation, a step that is not valid in general. However, one can show for all of the examples we will consider this is valid, so we will take the physicists' approach and ignore this detail in our calculations.

Kolmogorov's forward equation. This time we split $t + h$ up in a different way when we use the Chapman–Kolmogorov equations:

$$p_{t+h}(i,j) - p_t(i,j) = \left(\sum_{k} p_t(i,k)p_h(k,j)\right) - p_t(i,j)$$

$$= \left(\sum_{k \neq j} p_t(i,k)p_h(k,j)\right) + [p_h(j,j) - 1]p_t(i,j)$$

Computing as before we arrive at

(2.5) $$p'_t(i,j) = \sum_{k \neq j} p_t(i,k)q(k,j) - p_t(i,j)\lambda_j$$

Introducing matrix notation again, we can write

(2.6) $$p'_t = p_t Q$$

Comparing (2.6) with (2.3) we see that $p_t Q = Q p_t$ and that the two forms of Kolmogorov's differential equations correspond to writing the rate matrix on the left or the right. While we are on the subject of the choices, we should remember that in general for matrices $AB \neq BA$, so it is somewhat remarkable that $p_t Q = Q p_t$. The key to the fact that these matrices commute is that $p_t = e^{Qt}$ is made up of powers of Q:

$$Q \cdot e^{Qt} = \sum_{n=0}^{\infty} Q \cdot \frac{(Qt)^n}{n!} = \sum_{n=0}^{\infty} \frac{(Qt)^n}{n!} \cdot Q = e^{Qt} \cdot Q$$

To illustrate the use of Kolmogorov's equations we will now consider some examples. The simplest possible is

Example 2.1. Poisson process. Let $X(t)$ be the number of arrivals up to time t in a Poisson process with rate λ. In order to go from i arrivals at time s to j arrivals at time $t+s$ we must have $j \geq i$ and have exactly $j - i$ arrivals in t units of time, so

$$(2.7) \qquad p_t(i,j) = e^{-\lambda t} \frac{(\lambda t)^{j-i}}{(j-i)!}$$

To check the differential equation we have to first figure out what it is. Using the more explicit form of the backwards equation, (2.2), and plugging in our rates, we have

$$p'_t(i,j) = \lambda p_t(i+1,j) - \lambda p_t(i,j)$$

To check this we have to differentiate the formula in (2.7). When $j > i$ we have that the derivative of (2.7) is

$$-\lambda e^{-\lambda t} \frac{(\lambda t)^{j-i}}{(j-i)!} + e^{-\lambda t} \frac{(\lambda t)^{j-i-1}}{(j-i-1)!} \lambda = -\lambda p_t(i,j) + \lambda p_t(i+1,j)$$

When $j = i$, $p_t(i,i) = e^{-\lambda t}$, so the derivative is

$$-\lambda e^{-\lambda t} = -\lambda p_t(i,i) = -\lambda p_t(i,i) + \lambda p_t(i+1,i)$$

since $p_t(i+1,i) = 0$. □

There are not many examples in which one can write down solutions of the Kolmogorov's differential equation. A remarkable exception is:

Example 2.2. Yule process. In this model each particle splits into two at rate β, so $q(i, i+1) = \beta i$. To find the transition probability of the Yule process we will guess and verify that

$$(2.8) \qquad p_t(1,j) = e^{-\beta t}(1 - e^{-\beta t})^{j-1} \qquad \text{for } j \geq 1$$

i.e., a geometric distribution with success probability $e^{-\beta t}$.

Proof. To check this we will use the forward equation (2.5) to conclude

$$(*) \qquad p'_t(1,j) = -\beta j p_t(1,j) + \beta(j-1) p_t(1, j-1)$$

The use of the forward equation here is dictated by the fact that we are only writing down formulas for $p_t(1,j)$. To check $(*)$ we differentiate the formula proposed in (2.8) to see that if $j > 1$

$$\begin{aligned}p'_t(1,j) =& -\beta e^{-\beta t}(1 - e^{-\beta t})^{j-1} \\&+ e^{-\beta t}(j-1)(1-e^{-\beta t})^{j-2}(\beta e^{-\beta t})\end{aligned}$$

Recopying the first term on the right and using $\beta e^{-\beta t} = -(1 - e^{-\beta t})\beta + \beta$ in the second, we can rewrite the right-hand side of the above as

$$-\beta e^{-\beta t}(1 - e^{-\beta t})^{j-1} - e^{-\beta t}(j-1)(1 - e^{-\beta t})^{j-1}\beta$$
$$+ e^{-\beta t}(1 - e^{-\beta t})^{j-2}(j-1)\beta$$

Adding the first two terms then comparing with (*) shows that the above is

$$= -\beta j p_t(1, j) + \beta(j-1)p_t(1, j-1) \qquad \square$$

Having worked to find $p_t(1, j)$, it is fortunately easy to find $p_t(i, j)$. The chain starting with i individuals is the sum of i copies of the chain starting from 1 individual. Using this one can easily compute that

(2.9) $$p_t(i, j) = \binom{j-1}{i-1}(e^{-\beta t})^i(1 - e^{-\beta t})^{j-i}$$

In words, the sum of i geometrics has a negative binomial distribution.

Proof. To begin we note that if $N_1, \ldots N_i$ have the distribution given in (2.8) and $n_1 + \cdots + n_i = j$, then

$$P(N_1 = n_1, \ldots, N_i = n_i) = \prod_{k=1}^{i} e^{-\beta t}(1 - e^{-\beta t})^{n_k - 1} = (e^{-\beta t})^i(1 - e^{-\beta t})^{j-i}$$

To count the number of possible $(n_1, \ldots, n_i)$ with $n_k \geq 1$ and sum j, we think of putting j balls in a row. To divide the j balls into i groups of size $n_1, \ldots, n_i$, we will insert cards in the slots between the balls and let n_k be the number of balls in the kth group. Having made this transformation it is clear that the number of $(n_1, \ldots, n_i)$ is the number of ways of picking $i - 1$ of the $j - 1$ slot to put the cards or $\binom{j-1}{i-1}$. Multiplying this times the probability for each $(n_1, \ldots, n_i)$ gives the result. $\square$

It is usually not possible to explicitly solve Kolmogorov's equations to find the transition probability. To illustrate the difficulties involved we will now consider:

Example 2.3. Two-state chains. For concreteness, we can suppose that the state space is $\{1, 2\}$. In this case, there are only two flip rates $q(1, 2) = \lambda$ and $q(2, 1) = \mu$, so when we fill in the diagonal with minus the sum of the flip rates on that row we get

$$Q = \begin{pmatrix} -\lambda & \lambda \\ \mu & -\mu \end{pmatrix}$$

Writing out the backward equation in matrix form, (2.3), now we have

$$\begin{pmatrix} p'_t(1,1) & p'_t(1,2) \\ p'_t(2,1) & p'_t(2,2) \end{pmatrix} = \begin{pmatrix} -\lambda & \lambda \\ \mu & -\mu \end{pmatrix} \begin{pmatrix} p_t(1,1) & p_t(1,2) \\ p_t(2,1) & p_t(2,2) \end{pmatrix}$$

Doing the first column of matrix multiplication on the right, we have

$$p'_t(1,1) = -\lambda p_t(1,1) + \lambda p_t(2,1) = -\lambda(p_t(1,1) - p_t(2,1))$$
$$p'_t(2,1) = \mu p_t(1,1) - \mu p_t(2,1) = \mu(p_t(1,1) - p_t(2,1))$$

To solve this we guess that

(2.10)
$$p_t(1,1) = \frac{\mu}{\mu+\lambda} + \frac{\lambda}{\mu+\lambda}e^{-(\mu+\lambda)t}$$
$$p_t(2,1) = \frac{\mu}{\mu+\lambda} - \frac{\mu}{\mu+\lambda}e^{-(\mu+\lambda)t}$$

and verify that $p_t(1,1) - p_t(2,1) = e^{-(\mu+\lambda)t}$, so

$$p'_t(1,1) = -\lambda e^{-(\mu+\lambda)t} = -\lambda(p_t(1,1) - p_t(2,1))$$
$$p'_t(2,1) = \mu e^{-(\mu+\lambda)t} = \mu(p_t(1,1) - p_t(2,1))$$

To prepare for the developments in the next section note that the probability of being in state 1 converges exponentially fast to the equilibrium value $\mu/(\mu+\lambda)$.

4.3. Limiting Behavior

The study of the limiting behavior of continuous time Markov chains is simpler than the theory for discrete time chains, since the randomness of the exponential holding times implies that we don't have to worry about aperiodicity. We will first state the main convergence result that is a combination of the discrete-time convergence theorem, (4.5), and strong law, (4.8), from Chapter 1 and then explain the terms it uses.

(3.1) Theorem. *If a continuous-time Markov chain X_t is irreducible and has a stationary distribution π, then*

$$\lim_{t \to \infty} p_t(i,j) = \pi(j)$$

Furthermore if $r(j)$ is the reward we earn in state i and $\sum_j \pi(j)|r(j)| < \infty$, then as $t \to \infty$

$$\frac{1}{t}\int_0^t r(X_s)\,ds \to \sum_y \pi(y)r(y)$$

Here by X_t is **irreducible**, we mean that for any two states x and y it is possible to get from x to y in a finite number of jumps. To be precise, there is a sequence of states $x_0 = x, x_1, \ldots x_n = y$ so that $q(x_{m-1}, x_m) > 0$ for $1 \le m \le n$.

In discrete time a stationary distribution is a solution of $\pi p = \pi$. Since there is no first $t > 0$, in continuous time we need the stronger notion: π is said to be a **stationary distribution** if $\pi p_t = \pi$ for all $t > 0$. The last condition is difficult to check since it involves all of the p_t, and as we have seen in the previous section, the p_t are not easy to compute. The next result solves these problems by giving a test for stationarity in terms of the basic data used to describe the chain, the matrix of transition rates

$$Q(i,j) = \begin{cases} q(i,j) & j \ne i \\ -\lambda_i & j = i \end{cases}$$

where $\lambda_i = \sum_{j \ne i} q(i,j)$ is the total rate of transitions out of i.

(3.2) Theorem. π *is a stationary distribution if and only if* $\pi Q = 0$.

Why is this true? Filling in the definition of Q and rearranging, the condition $\pi Q = 0$ becomes

$$\sum_{k \ne j} \pi(k) q(k,j) = \pi(j) \lambda_j$$

If we think of $\pi(k)$ as the amount of sand at k, the right-hand side represents the rate at which sand leaves j, while the left gives the rate at which sand arrives at j. Thus, π will be a stationary distribution if for each j the flow of sand in to j is equal to the flow out of j.

More details. (2.6) says that $p'_t = p_t Q$. Multiplying by $\pi(i)$ and summing gives

($\star$) $$\sum_i \pi(i) p'_t(i,j) = \sum_{i,k} \pi(i) p_t(i,k) Q(k,j)$$

Taking the derivative outside the sum, and using $\pi p_t = \pi$, the expression on the left is

$$\frac{d}{dt}\left(\sum_i \pi(i) p_t(i,j)\right) = \frac{d}{dt} \pi(j) = 0$$

Doing the sum over i first on the right-hand side of ($\star$), we have

$$\sum_i \pi(i) p_t(i,k) = \pi(k)$$

so (⋆) becomes
$$0 = \sum_k \pi(k)Q(k,j) = (\pi Q)_j$$
where the right-hand side is the jth component of the vector πQ.

At this point we have shown that if π is stationary, then $\pi Q = 0$. To go in the other direction, we begin by interchanging the derivative and sum to conclude that
$$\frac{d}{dt}\left(\sum_i \pi(i)p_t(i,j)\right) = \sum_i \pi(i)p'_t(i,j)$$

To evaluate the right-hand side, we use the other Kolmogorov differential equation, (2.3). Multiplying it by $\pi(i)$ and summing gives
$$\frac{d}{dt}\sum_i \pi(i)p_t(i,j) = \sum_k (\pi Q)_k \, p_t(k,j) = 0$$

the last following from our assumption that $\pi Q = 0$. Since the derivative is 0, πp_t is constant and must always be equal to π its value at 0. □

We will now consider some examples. The simplest one was already covered in the last section.

Example 3.1. Two state chains. Suppose that the state space is $\{1,2\}$, $q(1,2) = \lambda$, and $q(2,1) = \mu$, where both rates are positive. The equations $\pi Q = 0$ can be written as
$$\begin{pmatrix} \pi_1 & \pi_2 \end{pmatrix} \begin{pmatrix} -\lambda & \lambda \\ \mu & -\mu \end{pmatrix} = \begin{pmatrix} 0 & 0 \end{pmatrix}$$

The first equation says $-\lambda \pi_1 + \mu \pi_2 = 0$. Taking into account that we must have $\pi_1 + \pi_2 = 1$, it follows that
$$\pi_1 = \frac{\mu}{\lambda + \mu} \quad \text{and} \quad \pi_2 = \frac{\lambda}{\lambda + \mu}$$
□

Perhaps the second simplest example is

Example 3.2. L.A. weather chain. There are three states: 1 = sunny, 2 = smoggy, 3 = rainy. The weather stays sunny for an exponentially distributed number of days with mean 3, then becomes smoggy. It stays smoggy for an exponentially distributed number of days with mean 4, then rain comes. The rain lasts for an exponentially distributed number of days with mean 1, then sunshine returns. Remembering that for an exponential the rate is 1 over the mean, the verbal description translates into the following Q-matrix

$$\begin{array}{c|ccc} & 1 & 2 & 3 \\ \hline 1 & -1/3 & 1/3 & 0 \\ 2 & 0 & -1/4 & 1/4 \\ 3 & 1 & 0 & -1 \end{array}$$

The relation $\pi Q = 0$ leads to three equations, the first two of which are

$$-\frac{1}{3}\pi(1) + \pi(3) = 0 \qquad \frac{1}{3}\pi(1) - \frac{1}{4}\pi(2) = 0$$

Setting $\pi(1) = c$, it follows from the first equation that $\pi(3) = c/3$ and from the second equation that $\pi(2) = 4c/3$. The sum of the three π's is $8c/3$, so $c = 3/8$ and we have

$$\pi(1) = 3/8, \quad \pi(2) = 4/8, \quad \pi(3) = 1/8$$

To check our answer, note that the weather cycles between sunny, smoggy, and rainy spending independent exponentially distributed amounts of time with means 3, 4, and 1, so the limiting fraction of time spent in each state is just the mean time spent in that state over the mean cycle time, 8. □

Detailed balance condition. Generalizing from discrete time we can formulate this condition as:

(3.3) $$\pi(k)q(k,j) = \pi(j)q(j,k) \quad \text{for all } j, k$$

The reason for interest in this concept is

(3.4) Theorem. *If (3.3) holds, then π is a stationary distribution.*

Why is this true? The detailed balance condition implies that the flows of sand between each pair of sites are balanced, which then implies that the net amount of sand flowing into each vertex is 0, i.e., $\pi Q = 0$.

Proof. Summing this over all $k \neq j$ and recalling the definition of λ_j gives

$$\sum_{k \neq j} \pi(k)q(k,j) = \pi(j) \sum_{k \neq j} q(j,k) = \pi(j)\lambda_j$$

Rearranging we have

$$(\pi Q)_j = \sum_{k \neq j} \pi(k)q(k,j) - \pi(j)\lambda_j = 0 \qquad \square$$

As in discrete time, (3.3) is much easier to check but does not always hold.

EXERCISE 3.1. Show that the stationary distribution for Example 3.2 does not satisfy the detailed balance condition (3.3).

For a large class of examples where the condition does hold:

Birth and death chains. Suppose that $S = \{0, 1, \ldots, N\}$ with

$$q(n, n+1) = \lambda_n \quad \text{for } 0 \leq n < N$$
$$q(n, n-1) = \mu_n \quad \text{for } 0 < n \leq N$$

Here λ_n represents the birth rate when there are n individuals in the system, and μ_n denotes the death rate in that case.

If we suppose that all the λ_n and μ_n listed above are positive then the birth and death chain is irreducible, and we can divide to write the detailed balance condition as

(3.5) $$\pi(n) = \frac{\lambda_{n-1}}{\mu_n} \pi(n-1)$$

Using this again we have $\pi(n-1) = (\lambda_{n-2}/\mu_{n-1})\pi(n-2)$ and it follows that

$$\pi(n) = \frac{\lambda_{n-1}}{\mu_n} \cdot \frac{\lambda_{n-2}}{\mu_{n-1}} \cdot \pi(n-2)$$

Repeating the last reasoning leads to

(3.6) $$\pi(n) = \frac{\lambda_{n-1} \cdot \lambda_{n-2} \cdots \lambda_0}{\mu_n \cdot \mu_{n-1} \cdots \mu_1} \pi(0)$$

To check this formula and help remember it, note that (i) there are n terms in the numerator and in the denominator, and (ii) if the state space was $\{0, 1, \ldots, n\}$, then $\mu_0 = 0$ and $\lambda_n = 0$, so these terms cannot appear in the formula.

To illustrate the use of (3.6) we consider two concrete examples.

Example 3.3. Barbershop. A barber can cut hair at rate 3, where the units are people per hour, i.e., each haircut requires an exponentially distributed amount of time with mean 20 minutes. Suppose customers arrive at times of a rate 2 Poisson process, but will leave if both chairs in the waiting room are full. (a) What fraction of time will both chairs be full? (b) In the long run, how many customers does the barber serve per hour?

Solution. We define our state to be the number of customers in the system, so $S = \{0, 1, 2, 3\}$. From the problem description it is clear that

$$q(i, i-1) = 3 \quad \text{for } i = 1, 2, 3$$
$$q(i, i+1) = 2 \quad \text{for } i = 0, 1, 2$$

The detailed balance conditions say

$$2\pi(0) = 3\pi(1), \quad 2\pi(1) = 3\pi(2), \quad 2\pi(2) = 3\pi(3)$$

Setting $\pi(0) = c$ and solving, we have

$$\pi(1) = \frac{2c}{3}, \quad \pi(2) = \frac{2}{3} \cdot \pi(1) = \frac{4c}{9}, \quad \pi(3) = \frac{2}{3} \cdot \pi(2) = \frac{8c}{27}$$

The sum of the π's is $(27 + 18 + 12 + 8)c/27 = 65c/27$, so $c = 27/65$ and

$$\pi(0) = 27/65, \quad \pi(1) = 18/65, \quad \pi(2) = 12/65, \quad \pi(3) = 8/65$$

From this we see that 8/65's of the time both chairs are full, so that fraction of the arrivals are lost and hence 57/65's or 87.7% of the customers enter service. Since the original arrival rate is 2, this means he serves an average of $114/65 = 1.754$ customers per hour. □

EXERCISE 3.2. How many customers per hour would he serve if there were only one chair in the waiting room?

Example 3.4. Machine repair model. A factory has three machines in use and one repairman. Suppose each machine works for an exponential amount of time with mean 60 days between breakdowns, but each breakdown requires an exponential repair time with mean 4 days. What is the long-run fraction of time all three machines are working?

Solution. Let X_t be the number of working machines. Since there is one repairman we have $q(i, i+1) = 1/4$ for $i = 0, 1, 2$. On the other hand, the failure rate is proportional to the number of machines working, so $q(i, i-1) = i/60$ for $i = 1, 2, 3$. Setting $\pi(0) = c$ and plugging into the recursion (3.5) gives

$$\pi(1) = \frac{\lambda_0}{\mu_1} \cdot \pi(0) = \frac{1/4}{1/60} \cdot c = 15c$$

$$\pi(2) = \frac{\lambda_1}{\mu_2} \cdot \pi(1) = \frac{1/4}{2/60} \cdot 15c = \frac{225c}{2}$$

$$\pi(3) = \frac{\lambda_2}{\mu_3} \cdot \pi(2) = \frac{1/4}{3/60} \cdot \frac{225c}{2} = \frac{1225c}{2}$$

Adding up the π's gives $(1225 + 225 + 30 + 2)c/2 = 1480c/2$ so $c = 2/1480$ and we have

$$\pi(3) = \frac{1225}{1480} \quad \pi(2) = \frac{225}{1480} \quad \pi(1) = \frac{30}{1480} \quad \pi(0) = \frac{2}{1480}$$

Thus in the long run all three machines are working $1225/1480 = .8277$ of the time. □

For our final example of the use of detailed balance we consider

Example 3.5. M/M/s queue with balking. A bank has s tellers that serve customers who need an exponential amount of service with rate μ and queue in a single line if all of the servers are busy. Customers arrive at times of a Poisson process with rate λ but only join the queue with probability a_n if there are n customers in line. As noted in Example 1.4, the birth rate $\lambda_n = \lambda a_n$ for $n \geq 0$, while the death rate is

$$\mu_n = \begin{cases} n\mu & 0 \leq n \leq s \\ s\mu & n \geq s \end{cases}$$

for $n \geq 1$. It is reasonable to assume that if the line is long the probability the customer will join the queue is small. The next result shows that this is always enough to prevent the queue length from growing out of control.

(3.7) If $a_n \to 0$ as $n \to \infty$, then there is a stationary distribution.

Proof. It follows from (3.4) that if $n \geq s$, then

$$\pi(n+1) = \frac{\lambda_n}{\mu_{n+1}} \cdot \pi(n) = a_n \cdot \frac{\lambda}{s\mu} \cdot \pi(n)$$

If N is large enough and $n \geq N$, then $a_n \lambda/(s\mu) \leq 1/2$ and it follows that

$$\pi(n+1) \leq \frac{1}{2}\pi(n) \ldots \leq \left(\frac{1}{2}\right)^{n-N} \pi(N)$$

This implies that $\sum_n \pi(n) < \infty$, so we can pick $\pi(0)$ to make the sum $= 1$. □

Concrete example. Suppose $s = 1$ and $a_n = 1/(n+1)$. In this case

$$\frac{\lambda_{n-1} \cdots \lambda_0}{\mu_n \cdots \mu_1} = \frac{\lambda^n}{\mu^n} \cdot \frac{1}{1 \cdot 2 \cdots n} = \frac{(\lambda/\mu)^n}{n!}$$

176 Chapter 4 Continuous-Time Markov Chains

To find the stationary distribution we want to take $\pi(0) = c$ so that

$$c \sum_{n=0}^{\infty} \frac{(\lambda/\mu)^n}{n!} = 1$$

Recalling the formula for the Poisson distribution with mean λ/μ, we see that $c = e^{-\lambda/\mu}$ and the stationary distribution is Poisson. □

4.4. Queueing Chains

In this section we will take a systematic look at the basic models of queueing theory that have Poisson arrivals and exponential service times. The arguments concerning Wendy's in Section 3.2 explain why we can be happy assuming that the arrival process is Poisson. However, the assumption of exponential services times is hard to justify. Here, it is a necessary evil. The lack of memory property of the exponential is needed for the queue length to be a continuous Markov chain. Later, in Section 5.3 we will be able to prove some of these results for general service times. However the theory of queues is the most complete when the interarrival times and the service times are exponential.

We begin with the simplest example:

Example 4.1. M/M/1 queue. In this system customers arrive to a single server facility at the times of a Poisson process with rate λ, and each requires an independent amount of service that has an exponential distribution with rate μ. From the description it should be clear that the transition rates are

$$q(n, n+1) = \lambda \quad \text{if } n \geq 0$$
$$q(n, n-1) = \mu \quad \text{if } n \geq 1$$

so we have a birth and death chain with birth rates $\lambda_n = \lambda$ and death rates $\mu_n = \mu$. Plugging into our formula for the stationary distribution, (3.6), we have

(4.1) $$\pi(n) = \frac{\lambda_{n-1} \cdots \lambda_0}{\mu_n \cdots \mu_1} \cdot \pi(0) = \left(\frac{\lambda}{\mu}\right)^n \pi(0)$$

To find the value of $\pi(0)$, we recall that for $|\theta| < 1$, $\sum_{n=0}^{\infty} \theta^n = 1/(1-\theta)$. From this we see that if $\lambda < \mu$, then

$$\sum_{n=0}^{\infty} \pi(n) = \sum_{n=0}^{\infty} \left(\frac{\lambda}{\mu}\right)^n \pi(0) = \frac{\pi(0)}{1 - (\lambda/\mu)}$$

So to have the sum 1, we pick $\pi(0) = 1 - (\lambda/\mu)$, and the resulting stationary distribution is the shifted geometric distribution

(4.2) $$\pi(n) = \left(1 - \frac{\lambda}{\mu}\right)\left(\frac{\lambda}{\mu}\right)^n \quad \text{for } n \geq 0$$

Having determined the stationary distribution we can now compute various quantities of interest concerning the queue. We might be interested, for example, in the distribution of the waiting time W of a customer who arrives to find the queue in equilibrium. To do this we begin by noting that the only way to wait 0 is for the number of people waiting in the queue Q to be 0 so

$$P(W = 0) = P(Q = 0) = 1 - \frac{\lambda}{\mu}$$

When there is at least one person in the system, the arriving customer will spend a positive amount of time in the queue. Writing $P(W = x)$ for the density function of W on $(0, \infty)$, we note that if there are n people in the system when the customer arrives, then the amount of time he needs to enter service has a gamma(n, μ) density, so using (1.11) in Chapter 3

$$P(W = x) = \sum_{n=1}^{\infty} \left(1 - \frac{\lambda}{\mu}\right)\left(\frac{\lambda}{\mu}\right)^n e^{-\mu x} \frac{\mu^n x^{n-1}}{(n-1)!}$$

Changing variables $m = n - 1$ and rearranging, the above becomes

$$= \left(1 - \frac{\lambda}{\mu}\right) e^{-\mu x} \lambda \sum_{m=0}^{\infty} \frac{\lambda^m x^m}{m!} = \frac{\lambda}{\mu}(\mu - \lambda) e^{-(\mu - \lambda)x}$$

Recalling that $P(W > 0) = \lambda/\mu$, we can see that the last result says that the conditional distribution of W given that $W > 0$ is exponential with rate $\mu - \lambda$.

Example 4.2. M/M/1 queue with a finite waiting room. In this system customers arrive at the times of a Poisson process with rate λ. Customers enter service if there are $< N$ individuals in the system, but when there are $\geq N$ customers in the system, the new arrival leaves never to return. Once in the system, each customer requires an independent amount of service that has an exponential distribution with rate μ.

Taking the state to be the number of customers in the system, the state space is now $S = \{0, 1, \ldots N\}$. The birth and death rates are changed a little

$$q(n, n+1) = \lambda \quad \text{if } 0 \leq n < N$$
$$q(n, n-1) = \mu \quad \text{if } 0 < n \leq N$$

but our formula for the stationary distribution, (3.6), still gives

$$\pi(n) = \frac{\lambda_{n-1} \cdots \lambda_0}{\mu_n \cdots \mu_1} \cdot \pi(0) = \left(\frac{\lambda}{\mu}\right)^n \pi(0) \qquad \text{for } 1 \leq n \leq N$$

The first thing that changes in the analysis is the normalizing constant. To isolate the arithmetic from the rest of the problem we recall that if $\theta \neq 1$, then

(4.3) $$\sum_{n=0}^{N} \theta^n = \frac{1 - \theta^{N+1}}{1 - \theta}$$

Suppose now that $\lambda \neq \mu$. Using (4.3), we see that if

$$c = \frac{1 - \lambda/\mu}{1 - (\lambda/\mu)^{N+1}}$$

then the sum is 1, so the stationary distribution is given by

(4.4) $$\pi(n) = \frac{1 - \lambda/\mu}{1 - (\lambda/\mu)^{N+1}} \left(\frac{\lambda}{\mu}\right)^n \qquad \text{for } 0 \leq n \leq N$$

The new formula is similar to the old one in (4.2) and when $\lambda < \mu$ reduces to it as $N \to \infty$. Of course, when the waiting room is finite, the state space is finite and we always have a stationary distribution, even when $\lambda > \mu$. The analysis above has been restricted to $\lambda \neq \mu$. However, it is easy to see that when $\lambda = \mu$ the stationary distribution is $\pi(n) = 1/(N+1)$ for $0 \leq n \leq N$.

To check formula (4.4), we note that the barbershop chain, Example 3.3, has this form with $N = 3$, $\lambda = 2$, and $\mu = 3$, so plugging into (4.4) and multiplying numerator and denominator by $3^4 = 81$, we have

$$\pi(0) = \frac{1 - 2/3}{1 - (2/3)^4} = \frac{81 - 54}{81 - 16} = 27/65$$

$$\pi(1) = \frac{2}{3}\pi(0) = 18/65$$

$$\pi(2) = \frac{2}{3}\pi(1) = 12/65$$

$$\pi(3) = \frac{2}{3}\pi(2) = 8/65$$

From a single server with a finite waiting room we move now to s servers with an unlimited waiting room, a system described more fully in Example 1.3.

Example 4.3. M/M/s queue. Imagine a bank with $s \geq 1$ tellers that serve customers who queue in a single line if all servers are busy. We imagine that customers arrive at the times of a Poisson process with rate λ, and each requires an independent amount of service that has an exponential distribution with rate μ. As explained in Example 1.3, the flip rates are $q(n, n+1) = \lambda$ and

$$q(n, n-1) = \begin{cases} \mu n & \text{if } n \leq s \\ \mu s & \text{if } n \geq s \end{cases}$$

The conditions that result from using the detailed balance condition are

(4.5) $\quad \lambda \pi(0) = \mu \pi(1) \quad \lambda \pi(1) = 2\mu \pi(2) \quad \ldots \quad \lambda \pi(s-1) = s\mu \pi(s)$

Then for $k \geq 0$ we have

$$\pi(s+k-1)\lambda = s\mu \pi(s+k) \quad \text{or} \quad \pi(s+k) = \frac{\lambda}{s\mu} \pi(s+k-1)$$

Iterating the last equation, we have that for $k \geq 0$

(4.6) $\quad \pi(s+k) = \left(\frac{\lambda}{s\mu}\right)^2 \pi(s+k-2) \ldots = \left(\frac{\lambda}{s\mu}\right)^{k+1} \pi(s-1)$

From the last formula we see that if $\lambda < s\mu$ then $\sum_{k=0}^{\infty} \pi(s+k) < \infty$ so $\sum_{j=0}^{\infty} \pi(j) < \infty$ and it is possible to pick $\pi(0)$ to make the sum equal to 1. From this it follows that

(4.7) $\quad$ If $\lambda < s\mu$, then the M/M/s queue has as stationary distribution.

The condition $\lambda < s\mu$ for the existence of a stationary distribution is natural since it says that the service rate of the fully loaded system is larger than the arrival rate, so the queue will not grow out of control. Conversely,

(4.8) $\quad$ If $\lambda > s\mu$, the M/M/s queue is transient.

Why is this true? The conclusion comes from combining two ideas:

(i) An M/M/s queue with s rate μ servers is less efficient than an M/M/1 queue with 1 rate $s\mu$ server, since the single server queue always has departures at rate $s\mu$, while the s server queue sometimes has departures at rate $n\mu$ with $n < s$.

(ii) An M/M/1 queue is transient if its arrival rate is larger than its service rate. $\square$

180   Chapter 4   Continuous-Time Markov Chains

Formulas for the stationary distribution $\pi(n)$ for the $M/M/s$ queue are unpleasant to write down for a general number of servers s, but it is not hard to use (4.5) and (4.6) to find the stationary distribution in concrete cases:

Example 4.4. M/M/3 queue. When $s = 3$, $\lambda = 2$ and $\mu = 1$, the first two equations in (4.5) say

$$(\star) \qquad 2\pi(0) = \pi(1) \qquad 2\pi(1) = 2\pi(2)$$

while (4.6) tells us that for $k \geq 0$

$$\pi(3+k) = \left(\frac{2}{3}\right)^{k+1} \pi(2)$$

Summing the last result from $k = 0$ to ∞, adding $\pi(2)$, and changing variables $j = k+1$, we have

$$\sum_{m=2}^{\infty} \pi(m) = \pi(2) \sum_{j=0}^{\infty} (2/3)^j = \frac{\pi(2)}{1-\frac{2}{3}} = 3\pi(2)$$

by the formula for the sum of the geometric series. Setting $\pi(2) = c$ and using $(\star)$, we see that

$$\pi(1) = c \qquad \pi(0) = \frac{1}{2}\pi(1) = c/2$$

Taking the contributions in order of increasing j, the sum of all the $\pi(j)$ is $(0.5 + 1 + 3)\pi(2)$. From this we conclude that $\pi(2) = 2/9$, so

$$\pi(0) = 1/9, \quad \pi(1) = 2/9, \quad \pi(k) = (2/9)(2/3)^{k-2} \text{ for } k \geq 2$$

Example 4.5. M/M/∞ queue. In this system customers arrive at the times of a Poisson process with rate λ, and each requires an independent amount of service that has an exponential distribution with rate μ. However, this time there are infinitely many servers, so (assuming we start with finitely many customers in the system!) there is never a queue and each customer can immediately enter service.

A system with infinitely many servers may sound absurd, but this model is useful for studying telephone traffic. To determine the number of lines that are needed for Cornell employees to call off-campus, we can pretend that there are always enough lines (i.e., infinitely many), find the stationary distribution, and then use this to estimate how many lines are needed so that 99% (or 99.99%) of the time there will be enough capacity.

From the description of the queueing system it should be clear that the transition rates are

$$q(n, n+1) = \lambda \quad \text{if } n \geq 0$$
$$q(n, n-1) = \mu n \quad \text{if } n \geq 1$$

Plugging into our formula for the stationary distribution, (3.6), we have

$$\pi(n) = \frac{\lambda_{n-1} \cdots \lambda_0}{\mu_n \cdots \mu_1} \cdot \pi(0)$$
$$= \frac{\lambda^n}{\mu^n \cdot n \cdot (n-1) \cdots 2 \cdot 1} \cdot \pi(0) = \frac{(\lambda/\mu)^n}{n!} \cdot \pi(0)$$

To determine the value of $\pi(0)$ to make the sum 1, we recall that if X has a Poisson distribution with mean θ, then

$$P(X = n) = e^{-\theta} \frac{\theta^n}{n!}$$

So if $\theta = \lambda/\mu$ and $\pi(0) = e^{-\lambda/\mu}$, then

$$\pi(n) = e^{-\lambda/\mu} \frac{(\lambda/\mu)^n}{n!} \quad \text{for } n \geq 0$$

i.e., the stationary distribution is Poisson with mean λ/μ.

4.5. Reversibility

Our main aim in this section is to prove the following remarkable fact:

(5.1) Theorem. *If $\lambda < \mu s$, then the output process of the M/M/s queue in equilibrium is a rate λ Poisson process.*

Your first reaction to this should be that it is crazy. Customers depart at rate $0, \mu, 2\mu, \ldots, s\mu$, depending on the number of servers that are busy and it is usually the case that none of these numbers $= \lambda$. To further emphasize the surprising nature of (5.1), suppose for concreteness that there is one server, $\lambda = 1$, and $\mu = 10$. If, in this situation, we have just seen 30 departures in the last 2 hours, then it seems reasonable to guess that the server is busy and the next departure will be exponential(10). However, if the output process is Poisson, then the number of departures in disjoint intervals are independent.

Our first step in making the result in (5.1) seem reasonable is to check by hand that if there is one server and the queue is in equilibrium, then the time

Chapter 4 Continuous-Time Markov Chains

of the first departure, D_1, has an exponential distribution with rate λ. There are two cases to consider.

Case 1. If there are $n \geq 1$ customers in the queue, then the time to the next departure has an exponential distribution with rate μ.

Case 2. If there are $n = 0$ customers in the queue, then we have to wait an exponential(λ) amount of time until the first arrival, and then an independent exponential(μ) for that customer to depart. If we let T_1 and T_2 be the waiting times for the arrival and for the departure, then breaking things down according to the value of T_1,

$$P(T_1 + T_2 = t) = \int_0^t P(T_1 = s) P(T_2 = t - s) \, ds$$

$$= \int_0^t \lambda e^{-\lambda s} \cdot \mu e^{-\mu(t-s)} \, ds = \lambda \mu e^{-\mu t} \int_0^t e^{-(\lambda - \mu)s} \, ds$$

$$= \frac{\lambda \mu e^{-\mu t}}{\lambda - \mu} \left(1 - e^{-(\lambda - \mu)t}\right) = \frac{\lambda \mu}{\lambda - \mu} \left(e^{-\mu t} - e^{-\lambda t}\right)$$

The probability of 0 customers is $1 - (\lambda/\mu)$ by (4.2). This implies the probability of ≥ 1 customer is λ/μ, so combining the two cases:

$$P(D_1 = t) = \frac{\mu - \lambda}{\mu} \cdot \frac{\lambda \mu}{\lambda - \mu} \left(e^{-\mu t} - e^{-\lambda t}\right) + \frac{\lambda}{\mu} \cdot \mu e^{-\mu t}$$

At this point cancellations occur to produce the answer we claimed:

$$-\lambda \left(e^{-\mu t} - e^{-\lambda t}\right) + \lambda e^{-\mu t} = \lambda e^{-\lambda t}$$

We leave it to the adventurous reader to try to repeat the last calculation for the M/M/s queue with $s > 1$ where there is not a neat formula for the stationary distribution.

To begin to develop the theory we need to prove (5.1), we return to discrete time. Let $p(i, j)$ be a transition probability with stationary distribution $\pi(i)$. Let X_n be a realization of the Markov chain starting from the stationary distribution, i.e., $P(X_0 = i) = \pi(i)$. The next result says that if we watch the process X_m, $0 \leq m \leq n$, backwards, then it is a Markov chain.

(5.2) Lemma. *Fix n and let $Y_m = X_{n-m}$ for $0 \leq m \leq n$. Then Y_m is a Markov chain with transition probability*

$$r(i, j) = P(Y_{m+1} = j | Y_m = i) = \frac{\pi(j) p(j, i)}{\pi(i)}$$

Section 4.5 Reversibility

Proof. We need to calculate the conditional probability.

$$P(Y_{m+1} = i_{m+1} | Y_m = i_m, Y_{m-1} = i_{m-1} \ldots Y_0 = i_0)$$
$$= \frac{P(X_{n-(m+1)} = i_{m+1}, X_{n-m} = i_m, X_{n-m+1} = i_{m-1} \ldots X_n = i_0)}{P(X_{n-m} = i_m, X_{n-m+1} = i_{m-1} \ldots X_n = i_0)}$$

Using the Markov property, we see the numerator is equal to

$$\pi(i_{m+1}) p(i_{m+1}, i_m) P(X_{n-m+1} = i_{m-1}, \ldots X_n = i_0 | X_{n-m} = i_m)$$

Similarly the denominator can be written as

$$\pi(i_m) P(X_{n-m+1} = i_{m-1}, \ldots X_n = i_0 | X_{n-m} = i_m)$$

Dividing the last two formulas and noticing that the conditional probabilities cancel we have

$$P(Y_{m+1} = i_{m+1} | Y_m = i_m, \ldots Y_0 = i_0) = \frac{\pi(i_{m+1}) p(i_{m+1}, i_m)}{\pi(i_m)}$$

This shows Y_m is a Markov chain with the indicated transition probability. □

The formula for the transition probability in (5.2) may look a little strange, but it is easy to see that it works; i.e., the $r(i, j) \geq 0$, and have

$$\sum_j r(i,j) = \sum_j \frac{\pi(j) p(j,i)}{\pi(i)} = \frac{\pi(i)}{\pi(i)} = 1$$

since $\pi p = \pi$. When π satisfies the detailed balance conditions:

$$\pi(i) p(i, j) = \pi(j) p(j, i)$$

the transition probability for the reversed chain,

$$r(i,j) = \frac{\pi(j) p(j,i)}{\pi(i)} = p(i,j)$$

is the same as the original chain. In words, if we make a movie of the Markov chain X_m, $0 \leq m \leq n$ starting from an initial distribution that satisfies the detailed balance condition and watch it backwards (i.e., consider $Y_m = X_{n-m}$ for $0 \leq m \leq n$), then we see a random process with the same distribution.

Continuous time. Up to this point we have only considered discrete time. In continuous time the details of the proofs are a little harder to write down but the facts are the same. Let X_t be a realization of the Markov chain with jump rates $q(i,j)$ and transition probability $p_t(i,j)$ starting from the stationary distribution, i.e., $P(X_0 = i) = \pi(i)$.

(5.3) Lemma. Fix t and let $Y_s = X_{t-s}$ for $0 \le s \le t$. Then Y_s is a Markov chain with transition probability

$$r_t(i,j) = \frac{\pi(j)p_t(j,i)}{\pi(i)}$$

If π satisfies the detailed balance condition $\pi(i)q(i,j) = \pi(j)q(j,i)$, then the reversed chain has transition probability $r_t(i,j) = p_t(i,j)$.

With (5.3) established, we are ready to give the

Proof of (5.1). As we learned in Example 4.3, when $\lambda < \mu s$ the $M/M/s$ queue is a birth and death chain with a stationary distribution π that satisfies the detailed balance condition. (5.3) implies that if we take the movie of the Markov chain in equilibrium then we see something that has the same distribution as the $M/M/s$ queue. Reversing time turns arrivals into departures, so the departures must be a Poisson process with rate λ. □

It should be clear from the proof just given that we also have:

(5.4) Theorem. Consider a queue in which arrivals occur according to a Poisson process with rate λ and customers are served at rate μ_n when there are n in the system. Then as along as there is a stationary distribution the output process will be a rate λ Poisson process.

A second refinement of (5.1) that will be useful in the next section is

(5.5) Theorem. Let $N(t)$ be the number of departures between time 0 and time n for the $M/M/1$ queue $X(t)$ started from its equilibrium distribution. Then $\{N(s) : 0 \le s \le t\}$ and $X(t)$ are independent.

Why is this true? At first it may sound deranged to claim that the output process up to time t is independent of the queue length. However, if we reverse time, then the departures before time t turn into arrivals after t, and these are obviously independent of the queue length at time t, $X(t)$. □

4.6. Queueing Networks

In many situations we are confronted with more than one queue. For example, in California when you go to the Department of Motor Vehicles to renew your driver's license you must (i) take a test on the driving laws, (ii) have your test graded, (iii) pay your fees, and (iv) get your picture taken. A simple model of this type of situation with only two steps is:

Example 6.1. Two-station tandem queue. In this system customers at times of a Poisson process with rate λ arrive at service facility 1 where they each require an independent exponential amount of service with rate μ_1. When they complete service at the first site, they join a second queue to wait for an exponential amount of service with rate μ_2.

Our main problem is to find conditions that guarantee that the queue stabilizes, i.e., has a stationary distribution. This is simple in the tandem queue. The first queue is not affected by the second, so if $\lambda < \mu_1$, then (4.2) tells us that the equilibrium probability of the number of customers in the first queue, N_t^1, is given by the shifted geometric distribution

$$P(N_t^1 = m) = \left(\frac{\lambda}{\mu_1}\right)^m \left(1 - \frac{\lambda}{\mu_1}\right)$$

In the previous section we learned that the output process of an $M/M/1$ queue in equilibrium is a rate λ Poisson process. This means that if the first queue is in equilibrium, then the number of customers in the queue, N_t^2, is itself an $M/M/1$ queue with arrivals at rate λ (the output rate for 1) and service rate μ_2. Using the results in (4.2) again, the number of individuals in the second queue has stationary distribution

$$P(N_t^2 = n) = \left(\frac{\lambda}{\mu_2}\right)^n \left(1 - \frac{\lambda}{\mu_2}\right)$$

To specify the stationary distribution of the system, we need to know the joint distribution of N_t^1 and N_t^2. The answer is somewhat remarkable: in equilibrium the two queue lengths are independent.

(6.1) $\qquad P(N_t^1 = m, N_t^2 = n) = \left(\frac{\lambda}{\mu_1}\right)^m \left(1 - \frac{\lambda}{\mu_1}\right) \cdot \left(\frac{\lambda}{\mu_2}\right)^n \left(1 - \frac{\lambda}{\mu_2}\right)$

Why is this true? The easiest way to see this is to use (5.5) to conclude that if we start the first queue in equilibrium, then its departure process up to time t is independent of the number of people in the its queue at time t, N_t^1. Since

186 Chapter 4 Continuous-Time Markov Chains

the number of people in the second queue is determined by its arrival process (which is the departure process from the first queue) and independent random variables, it follows that if we start with the first queue in equilibrium and an independent number of individuals in the second queues at time 0, then we have this property at all times $t > 0$ and hence in equilibrium. □

Since (5.5) was based on the magic of reversibility and in the last argument there is more than a little hand-waving going on, it is comforting to note that one can simply verify from the definitions that

(6.2) *If $\pi(m,n) = c\lambda^{m+n}/(\mu_1^m \mu_2^n)$, where $c = (1 - \lambda/\mu_1)(1 - \lambda/\mu_2)$ is a constant chosen to make the probabilities sum to 1, then π is a stationary distribution.*

Proof. The first step in checking $\pi Q = 0$ is to compute the rate matrix Q. To do this it is useful to draw a picture.

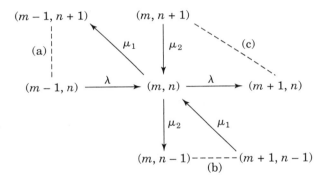

Here we have assumed that m and n are both positive. The rate arrows plus the dotted lines on the picture, make three triangles. We will now check that the flows out of and into (m,n) in each triangle balance. In symbols we note that

(a) $$\lambda \pi(m-1, n) = \frac{c\lambda^{m+n}}{\mu_1^{m-1} \mu_2^n} = \mu_1 \pi(m, n)$$

(b) $$\mu_1 \pi(m+1, n-1) = \frac{c\lambda^{m+n}}{\mu_1^m \mu_2^{n-1}} = \mu_2 \pi(m, n)$$

(c) $$\mu_2 \pi(m, n+1) = \frac{c\lambda^{m+n+1}}{\mu_1^m \mu_2^n} = \lambda \pi(m, n)$$

This shows that $\pi Q = 0$ when $m, n > 0$. There are three other cases to consider: (i) $m = 0$, $n > 0$, (ii) $m > 0$, $n = 0$, and (iii) $m = 0$, $n = 0$. For (i) we note that the rates in (a) are missing but since those in (b) and in (c) each balance, $\pi Q = 0$. For (ii) we note that the rates in (b) are missing but since those in (a) and in (c) each balance, $\pi Q = 0$. For (iii) we note that the rates in (a) and (b) are missing but since those in (c) each balance, $\pi Q = 0$. □

Example 6.2. General two-station queue. Suppose that at station i: arrivals from outside the system occur at rate λ_i, service occurs at rate μ_i, and departures go to the other queue with probability p_i and leave the system with probability $1 - p_i$.

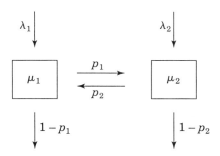

Our question is: When is the system stable? That is, when is there a stationary distribution? To get started on this question suppose that both servers are busy. In this case work arrives at station 1 at rate $\lambda_1 + p_2\mu_2$, and work arrives at station 2 at rate $\lambda_2 + p_1\mu_1$. It should be intuitively clear that:

(i) if $\lambda_1 + p_2\mu_2 < \mu_1$ and $\lambda_2 + p_1\mu_1 < \mu_2$, then each server can handle their maximum arrival rate and the system will have a stationary distribution.

(ii) if $\lambda_1 + p_2\mu_2 > \mu_1$ and $\lambda_2 + p_1\mu_1 > \mu_2$, then there is positive probability that both servers will stay busy for all time and the queue lengths will tend to infinity.

Not covered by (i) or (ii) is the situation in which server 1 can handle her worst case scenario but server 2 cannot cope with his:

$$\lambda_1 + p_2\mu_2 < \mu_1 \quad \text{and} \quad \lambda_2 + p_1\mu_1 > \mu_2$$

In some situations in this case, queue 1 will be empty often enough to reduce the arrivals at station 2 so that server 2 can cope with his workload. As we will see, a concrete example of this phenomenon occurs when

$$\lambda_1 = 1, \quad \mu_1 = 4, \quad p_1 = 1/2 \quad \lambda_2 = 2, \quad \mu_2 = 3.5, \quad p_2 = 1/4$$

To check that for these rates server 1 can handle the maximum arrival rate but server 2 cannot, we note that

$$\lambda_1 + p_2\mu_2 = 1 + \frac{1}{4} \cdot 3.5 = 1.875 < 4 = \mu_1$$

$$\lambda_2 + p_1\mu_1 = 2 + \frac{1}{2} \cdot 4 = 4 > 3.5 = \mu_2$$

To derive general conditions that will allow us to determine when a two-station network is stable, let r_i be the long-run average rate that customers arrive at station i. If there is a stationary distribution, then r_i must also be the long run average rate at which customers leave station i or the queue would grow linearly in time. If we want the flow in and out of each of the stations to balance, then we need

(6.3) $$r_1 = \lambda_1 + p_2 r_2 \quad \text{and} \quad r_2 = \lambda_2 + p_1 r_1$$

Plugging in the values for this example and solving gives

$$r_1 = 1 + \frac{1}{4}r_2 \quad \text{and} \quad r_2 = 2 + \frac{1}{2}r_1 = 2 + \frac{1}{2}\left(1 + \frac{1}{4}r_2\right)$$

So $(7/8)r_2 = 5/2$ or $r_2 = 20/7$, and $r_1 = 1 + 20/28 = 11/7$. Since

$$r_1 = 11/7 < 3 = \mu_1 \quad \text{and} \quad r_2 = 20/7 < 3.5$$

this analysis suggests that there will be a stationary distribution.

To prove that there is one, we return to the general situation and suppose that the r_i we find from solving (6.1) satisfy $r_i < \mu_i$. Thinking of two independent $M/M/1$ queues with arrival rates r_i, we let $\alpha_i = r_i/\mu_i$ and guess:

(6.4) If $\pi(m,n) = c\alpha_1^m \alpha_2^n$ where $c = (1-\alpha_1)(1-\alpha_2)$ then π is a stationary distribution.

Proof. The first step in checking $\pi Q = 0$ is to compute the rate matrix Q. To do this it is useful to draw a picture. Here, we have assumed that m and n are both positive. To make the picture slightly less cluttered, we have only labeled half of the arrows. Those that point in the same direction have the same rates.

Section 4.6 Queueing Networks

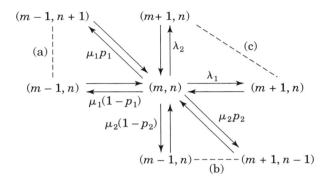

The rate arrows plus the dotted lines in the picture make three triangles. We will now check that the flows out of and into (m, n) in each triangle balance. In symbols we need to show that

(a) $\quad \mu_1 \pi(m,n) = \mu_2 p_2 \pi(m-1, n+1) + \lambda_1 \pi(m-1, n)$
(b) $\quad \mu_2 \pi(m,n) = \mu_1 p_1 \pi(m+1, n-1) + \lambda_2 \pi(m, n-1)$
(c) $\quad (\lambda_1 + \lambda_2) \pi(m,n) = \mu_2(1-p_2) \pi(m, n+1) + \mu_1(1-p_1) \pi(m+1, n)$

Filling in $\pi(m,n) = c\alpha_1^m \alpha_2^n$ and canceling out c, we have

$$\mu_1 \alpha_1^m \alpha_2^n = \mu_2 p_2 \alpha_1^{m-1} \alpha_2^{n+1} + \lambda_1 \alpha_1^{m-1} \alpha_2^n$$
$$\mu_2 \alpha_1^m \alpha_2^n = \mu_1 p_1 \alpha_1^{m+1} \alpha_2^{n-1} + \lambda_2 \alpha_1^m \alpha_2^{n-1}$$
$$(\lambda_1 + \lambda_2) \alpha_1^m \alpha_2^n = \mu_2(1-p_2) \alpha_1^m \alpha_2^{n+1} + \mu_1(1-p_1) \alpha_1^{m+1} \alpha_2^n$$

Canceling out the highest powers of α_1 and α_2 common to all terms in each equation gives

$$\mu_1 \alpha_1 = \mu_2 p_2 \alpha_2 + \lambda_1$$
$$\mu_2 \alpha_2 = \mu_1 p_1 \alpha_1 + \lambda_2$$
$$(\lambda_1 + \lambda_2) = \mu_2(1-p_2)\alpha_2 + \mu_1(1-p_1)\alpha_1$$

Filling in $\mu_i \alpha_i = r_i$, the three equations become

$$r_1 = p_2 r_2 + \lambda_1$$
$$r_2 = p_1 r_1 + \lambda_2$$
$$(\lambda_1 + \lambda_2) = r_2(1-p_2) + r_1(1-p_1)$$

The first two equations hold by (6.3). The third is the sum of the first two, so it holds as well.

Again this shows that $\pi Q = 0$ when $m, n > 0$ and there are three other cases to consider: (i) $m = 0$, $n > 0$, (ii) $m > 0$, $n = 0$, and (iii) $m = 0$, $n = 0$. For (i) we note that the rates in (a) are missing, but since those in (b) and in (c) each balance, $\pi Q = 0$. For (ii) we note that the rates in (b) are missing, but since those in (a) and in (c) each balance, $\pi Q = 0$. For (iii) we note that the rates in (a) and (b) are missing, but since those in (c) each balance, $\pi Q = 0$. □

Example 6.3. Network of M/M/1 queues. Assume now that there are stations $1 \leq i \leq K$. Arrivals from outside the system occur to station i at rate λ_i and service occurs there at rate μ_i. Departures go to station j with probability $p(i, j)$ and leave the system with probability

$$q(i) = 1 - \sum_j p(i, j)$$

To have a chance of stability we must suppose

(A) *For each i it is possible for a customer entering at i to leave the system. That is, for each i there is a sequence of states $i = j_0, j_1, \ldots j_n$ with $p(j_{m-1}, j_m) > 0$ for $1 \leq m \leq n$ and $q(j_n) > 0$.*

Generalizing (6.3), we investigate stability by solving the system of equations for the r_j that represent the arrival rate at station j. As remarked earlier, the departure rate from station j must equal the arrival rate, or a linearly growing queue would develop. Thinking about the arrival rate at j in two different ways, it follows that

(6.5) $$r_j = \lambda_j + \sum_{i=1}^{K} r_i p(i, j)$$

This equation can be rewritten in matrix form as $r = \lambda + rp$. In this form we can guess the solution

(6.6) $$r = \sum_{n=0}^{\infty} \lambda p^n = \sum_{n=0}^{\infty} \sum_{i=1}^{K} \lambda_i p^n(i, j)$$

where p^n denotes the nth power of the matrix, and check that it works

$$r = \lambda + \sum_{m=0}^{\infty} \lambda p^m \cdot p = \lambda + rp$$

The last calculation is informal but it can be shown that under assumption (A) the series defining r converges and our manipulations are justified. Putting aside these somewhat tedious details, it is easy to see that the answer in (6.6) is reasonable: $p^n(i, j)$ is the probability a customer entering at i is at j after he has completed n services. The sum then adds the rates for all the ways of arriving at j.

Having found the arrival rates at each station, we can again be brave and guess that if $r_j < \mu_j$, then the stationary distribution is given by

(6.7) $$\pi(n_1, \ldots, n_K) = \prod_{j=1}^{K} \left(\frac{r_j}{\mu_j}\right)^{n_j} \left(1 - \frac{r_j}{\mu_j}\right)$$

To prove this we will consider a more general collection of examples:

Example 6.4. Migration processes. As in the previous example, there are stations $1 \leq i \leq K$ and arrivals from outside the network occur to station i at rate λ_i. However, now when station i has n occupants, individuals depart at rate $\varphi_i(n)$ where $\varphi_i(n) \geq 0$ and $\varphi_i(0) = 0$. Finally, a customer leaving i goes to station j with probability $p(i, j)$ independent of past events. Our main motivation for considering this more general set-up is that by taking $1 \leq s_i \leq \infty$ and letting

$$\varphi_i(n) = \mu_i \min\{n, s_i\}$$

we can suppose that the ith station is an $M/M/s_i$ queue.

To find the stationary distribution for the migration process we first solve (6.5) to find the arrival and departure rates, r_i, for station i in equilibrium. Having done this we can let $\psi_i(n) = \prod_{m=1}^{n} \varphi_i(m)$ and introduce our second assumption:

(B) For $1 \leq j \leq K$, we have $\sum_{n=0}^{\infty} r_j^n / \psi_j(n) < \infty$

This condition guarantees that there is a constant $c_j > 0$ so that

$$\sum_{n=0}^{\infty} c_j r_j^n / \psi_j(n) = 1$$

The next result says that $\pi_j(n) = c_j r_j^n / \psi_j(n)$ gives the equilibrium probability that queue j has n individuals and that in equilibrium the queue lengths are independent.

(6.8) **Theorem.** *Suppose that conditions (A) and (B) hold. Then the migration process has stationary distribution*

$$\pi(n_1, \ldots, n_K) = \prod_{j=1}^{K} \frac{c_j r_j^{n_j}}{\psi_j(n_j)}$$

192 Chapter 4 Continuous-Time Markov Chains

To make the connection with the results for Example 6.3, note that if all the queues are single server, then $\psi_j(n) = \mu_j^n$, so (B) reduces to $r_j < \mu_j$ and when this holds the queue lengths are independent shifted geometrics.

We postpone the proof of (6.8) to the end of the section, since it is somewhat mysterious. To inspire the reader for the effort of reading the proof, we consider another example.

Example 6.5. Network of M/M/∞ queues. In this case $\varphi_j(n) = \mu_j n$, so

$$\psi_j(n) = \prod_{i=1}^{n} \varphi_j(m) = \mu_j^n \cdot n!$$

The constant defined in (B) has

$$1/c_j = \sum_{n=0}^{\infty} \frac{r_j^n}{\mu_j^n n!}$$

so by (6.8) the stationary distribution is

$$\pi(n_1, \ldots, n_K) = \prod_{j=1}^{K} = e^{-r_j/\mu_j} \frac{(r_j/\mu_j)^{n_j}}{n_j!}$$

i.e., the queue lengths are independent Poissons with mean r_j/μ_j.

One can also prove the last conclusion without using (6.8).

Direct proof. In a network of $M/M/\infty$ queues, there is no interference, so all individuals move independently between the stations. If the system starts empty at time 0, then the Poisson thinning result in (6.4) of Chapter 3 implies that the number of customers at j at time t will be independent Poissons with means

$$\int_0^t \sum_{i=1}^{K} \lambda_i p_s(i, j) \, ds$$

Here $p_t(i, j)$ is the transition probability for the chain that jumps at rate μ_i in state i, goes from i to j with probability $p(i, j)$, and leaves the state space to go to the absorbing state Δ with probability $q(i) = 1 - \sum_j p(i, j)$. Since each visit to j results in an exponentially distributed stay with mean $1/\mu_j$,

$$\int_0^\infty \lambda_i p_s(i, j) \, ds = \lambda_i \sum_{n=0}^{\infty} p^n(i, j) \frac{1}{\mu_j} = \frac{r_j}{\mu_j}$$

by the definition of r in (6.6). □

Proof of (6.8). Write n as shorthand for $(n_1, n_2, \ldots n_K)$. Let A_j (for arrival) be the operator that adds one customer to queue j, let D_j (for departure) be the operator that removes one from queue j, and let T_{jk} be the operator that transfers one customer from j to k. That is,

$$A_j n = \bar{n} \quad \text{where } \bar{n}_j = n_j + 1 \text{ and } \bar{n}_i = n_i \text{ for } i \neq j$$
$$D_j n = \bar{n} \quad \text{where } \bar{n}_j = n_j - 1 \text{ and } \bar{n}_i = n_i \text{ for } i \neq j$$

while $T_{jk} n = \bar{n}$ where $\bar{n}_j = n_j - 1$, $\bar{n}_k = n_k + 1$, and $\bar{n}_i = n_i$ otherwise. Note that if $n_j = 0$ then $D_j n$ and $T_{jk} n$ have -1 in the jth coordinate, so in this case $q(n, D_j n)$ and $q(n, T_{jk} n) = 0$.

In equilibrium the rate at which probability mass leaves n is the same as the rate at which it enters n. Taking into account the various ways the chain can leave or enter state n it follows that the condition for a stationary distribution $\pi Q = 0$ is equivalent to

$$\pi(n) \left(\sum_{k=1}^{K} q(n, A_k n) + \sum_{j=1}^{K} q(n, D_j n) + \sum_{j=1}^{K} \sum_{k=1}^{K} q(n, T_{jk} n) \right)$$
$$= \sum_{k=1}^{K} \pi(A_k n) q(A_k n, n)$$
$$+ \sum_{j=1}^{K} \pi(D_j n) q(D_j n, n) + \sum_{j=1}^{K} \sum_{k=1}^{K} \pi(T_{jk} n) q(T_{jk} n, n)$$

This will obviously be satisfied if we have

(6.9) $$\pi(n) \sum_{k=1}^{K} q(n, A_k n) = \sum_{k=1}^{K} \pi(A_k n) q(A_k n, n)$$

and for each j we have

(6.10) $$\pi(n) \left(q(n, D_j n) + \sum_{k=1}^{K} q(n, T_{jk} n) \right)$$
$$= \pi(D_j n) q(D_j n, n) + \sum_{k=1}^{K} \pi(T_{jk} n) q(T_{jk} n, n)$$

194 Chapter 4 Continuous-Time Markov Chains

Taking the second equation first, if $n_j = 0$, then both sides are 0, since $D_j n$ and $T_{jk} n$ are not in the state space. Supposing that $n_j > 0$ and filling in the values of our rates (6.10) becomes

(6.11)
$$\pi(n)\varphi_j(n_j)\left(q(j) + \sum_{k=1}^{K} p(j,k)\right)$$
$$= \pi(D_j n)\lambda_j + \sum_{k=1}^{K} \pi(T_{jk}n)\varphi_k(n_k+1)p(k,j)$$

The definition of q implies $q(j) + \sum_{k=1}^{K} p(j,k) = 1$, so filling in the proposed formula for $\pi(n)$ the left-hand side of (6.11) is

$$\pi(n)\varphi_j(n_j) = \prod_{i=1}^{K} \frac{c_i r_i^{n_i}}{\psi_i(n_i)} \cdot \varphi_j(n_j) = \prod_{i=1}^{K} \frac{c_i r_i^{\hat{n}_i}}{\psi_j(\hat{n}_i)} \cdot r_j = \pi(\hat{n}) \cdot r_j$$

where $\hat{n} = D_j n$ has $\hat{n}_j = n_j - 1$ and $\hat{n}_i = n_i$ for $i \neq j$. To compute the right-hand side of (6.11) we note that $(T_{jk}n)_i = \hat{n}_i$ for $i \neq k$ and $(T_{jk}n)_k = \hat{n}_k + 1 = n_k + 1$ so

$$\pi(T_{jk}n) = \pi(\hat{n}) \cdot \frac{r_k}{\varphi_k(n_k+1)}$$

Since $D_j n = \hat{n}$, we can rewrite the right-hand side of (6.11) as

$$= \pi(\hat{n})\lambda_j + \pi(\hat{n})\sum_{k=1}^{K} r_k p(k,j) = \pi(\hat{n}) \cdot r_j$$

where the last equality follows from (6.5): $\lambda_j + \sum_k r_k p(k,j) = r_j$.

At this point we have verified (6.10). Filling our rates into (6.9) and noting that $\pi(A_k n) = \pi(n) r_k / \varphi_k(n_k+1)$ we want to show

(6.12)
$$\pi(n) \sum_{k=1}^{K} \lambda_k = \pi(n) \sum_{k=1}^{K} \frac{r_k}{\varphi_k(n_k+1)} \cdot \varphi_k(n_k+1) q(k)$$

To derive this, we note that summing (6.5) from $j=1$ to K and interchanging the order of summation in the double sum on the right gives

$$\sum_{j=1}^{K} r_j = \sum_{j=1}^{K} \lambda_j + \sum_{k=1}^{K} r_k \sum_{j=1}^{K} p(k,j)$$
$$= \sum_{j=1}^{K} \lambda_j + \sum_{k=1}^{K} r_k - \sum_{k=1}^{K} r_k q(i)$$

since $\sum_{j=1}^{K} p(i,j) = 1 - q(i)$. Rearranging now gives

$$\sum_{k=1}^{K} r_k q(k) = \sum_{j=1}^{K} \lambda_j$$

This establishes (6.12), which implies (6.10), and completes the proof. □

4.7. Closed Queueing Networks

At first, the notion of N customers destined to move forever between K servers may sound like a queueing hell that might be the subject of a "Far Side" cartoon. However, as the next two examples show, this concept is useful for applications.

Example 7.1. Manufacturing system. The production of a part at a factory requires two operations. The first operation is always done at machine 1. The second is done at machine 2 or machine 3 with probabilities p and $1 - p$ after which the part leaves the system. Suppose that the factory has only a limited number of palettes, each of which holds one part. When a part leaves the system, the palette on which it rides is immediately used to bring a new part to queue at machine 1. If we ignore the parts, then the palettes are a closed queueing system. By computing the stationary distribution for this system we can compute the rate at which palettes leave machines 2 and 3, and hence compute the rate at which parts are made.

Example 7.2. Machine repair. For a concrete example consider trucks that can need engine repairs or tire repairs. To construct a closed queueing network model of this situation, we introduce three queues: $1 = $ the trucks that are working, $2 = $ those in engine repair, and $3 = $ those in tire repair. To simulate the breakdown mechanism, we will use an $M/M/\infty$ queue in which service times are exponential with rate λ if all trucks are always in use or an $M/M/s_1$ queue if there are never more than s_1 trucks in use. At repair station i we will suppose that there are s_i mechanics and hence an $M/M/s_i$ queue.

To begin to discuss these examples we need to introduce some notation and assumptions. Let $p(i,j)$ be the probability of going to station j after completing service at station i. We will suppose that

(A) *p is irreducible and has finitely many states*

so that there is a unique stationary distribution $\pi_i > 0$ for the routing matrix $p(i,j)$. If we let r_i denote the rate at which customers arrive at i in equilibrium

196   Chapter 4   Continuous-Time Markov Chains

then since there are no arrivals from outside, (6.6) becomes

$$(7.1) \qquad r_j = \sum_{i=1}^{K} r_i \, p(i,j)$$

If we divide r_j by $R = \sum_j r_j$, the result is a stationary distribution for p. However, the stationary distribution for p is unique so we must have $r_j = R\pi_j$. Note that the r_j are departure (or arrival) rates, so their sum, R, is not the number of customers in the system, but instead represents the average rate at which services are completed in the system. This can be viewed as a measure of the system's throughput rate. In Example 7.2, each part exits from exactly two queues, so $R/2$ is the rate at which new parts are made.

To find the stationary distribution for the queueing system we take a clue from Section 4.6 and guess that it will have a product form. This is somewhat of a crazy guess since there are a fixed total number of customers so the queues cannot possibly be independent. However, we will see that it works. As in the previous section we will again consider something more general than our queueing system.

Example 7.3. Closed migration processes are defined by three rules.

(a) No customers enter or leave the system.

(b) When there are n_i customers at station i departures occur at rate $\varphi_i(n_i)$ where $\varphi_i(0) = 0$.

(c) An individual leaving i goes to j with probability $p(i,j)$.

(7.2) Theorem. *Suppose (A) holds. The equilibrium distribution for the closed migration process with N individuals is*

$$\pi(n_1, \ldots, n_K) = c_N \prod_{i=1}^{K} \frac{\pi_i^{n_i}}{\psi_i(n_i)}$$

if $n_1 + \cdots + n_K = N$ and 0 otherwise. Here $\psi_i(n) = \prod_{m=1}^{n} \varphi_i(m)$ with $\psi_i(0) = 1$ and c_N is the constant needed to make the sum equal to 1.

Note. Here we used π_j instead of $r_j = R\pi_j$. Since $\sum_{j=1}^{K} n_j = N$, we have

$$\prod_{i=1}^{K} \frac{r_i^{n_i}}{\psi_i(n_i)} = R^N \prod_{i=1}^{K} \frac{\pi_i^{n_i}}{\psi_i(n_i)}$$

so this change only affects the value of the norming constant.

Section 4.7 Closed Queueing Networks

Proof. Write n as shorthand for $(n_1, n_2, \ldots n_K)$ and let T_{jk} be the operator that transfers one customer from j to k. That is, $T_{jk}n = \bar{n}$, where $\bar{n}_j = n_j - 1$, $\bar{n}_k = n_k + 1$, and $\bar{n}_i = n_i$ otherwise. The condition for a stationary distribution $\pi Q = 0$ is equivalent to

$$(7.3) \qquad \pi(n) \sum_{j=1}^{K} \sum_{k=1}^{K} q(n, T_{jk}n) = \sum_{j=1}^{K} \sum_{k=1}^{K} \pi(T_{jk}n) q(T_{jk}n, n)$$

This will obviously be satisfied if for each j we have

$$\pi(n) \sum_{k=1}^{K} q(n, T_{jk}n) = \sum_{k=1}^{K} \pi(T_{jk}n) q(T_{jk}n, n)$$

Filling in the values of our rates we want to show that

$$(7.4) \qquad \pi(n) \varphi_j(n_j) \sum_{k=1}^{K} p(j, k) = \sum_{k=1}^{K} \pi(T_{jk}n) \varphi_k(n_k + 1) p(k, j)$$

If $n_j = 0$, then both sides are 0, since $\varphi_j(0) = 0$ and $T_{jk}n$ is not in the state space. Thus we can suppose that $n_j > 0$. $\sum_{k=1}^{K} p(j, k) = 1$, so filling in the proposed value of $\pi(n)$, the left-hand side of (7.4) is

$$(7.5) \qquad \pi(n) \varphi_j(n_j) = c_N \prod_{i=1}^{K} \frac{\pi_i^{n_i}}{\psi_i(n_i)} \cdot \varphi_i(n_i) = \pi(\hat{n}) \cdot \pi_j$$

where $\hat{n}_j = n_j - 1$ and $\hat{n}_i = n_i$ otherwise. To compute the right-hand side of (7.4) we note that $(T_{jk}n)_i = \hat{n}_i$ for $i \neq k$ and $(T_{jk}n)_k = \hat{n}_k + 1 = n_k + 1$, so from the formula in (7.2),

$$\pi(T_{jk}n) = \pi(\hat{n}) \cdot \frac{\pi_k}{\varphi_k(n_k + 1)}$$

Using this we can rewrite:

$$(7.6) \qquad \sum_{k=1}^{K} \pi(T_{jk}n) \varphi_k(n_k + 1) p(k, j) = \pi(\hat{n}) \sum_{k=1}^{K} \pi_k p(k, j)$$

Since $\sum_k \pi_k p(k, j) = \pi_j$, the expressions in (7.5) and (7.6) are equal. This verifies (7.4) and we have proved the result. □

To see what (7.2) says we will now reconsider our two previous examples.

Example 7.4. Manufacturing system. Consider the special case of Example 7.1 in which the service rates are $\mu_1 = 2$, $\mu_2 = 1$, $\mu_3 = 1/3$, and the routing matrix is

$$p(i,j) = \begin{pmatrix} 0 & 2/3 & 1/3 \\ 1 & 0 & 0 \\ 1 & 0 & 0 \end{pmatrix}$$

To find the stationary distribution we set $\pi_1 = c$, then compute

$$\pi_2 = (2/3)c \quad \pi_3 = (1/3)c$$

The sum of the π's is $2c$, so $c = 1/2$. These queues are single servers so

$$\psi_i(n) = \prod_{m=1}^{n} \varphi_i(m) = \mu_i^n$$

Plugging into the formula in (7.2), we have

$$\pi(n_1, n_2, n_3) = c_N \prod_{i=1}^{K} \left(\frac{\pi_i}{\mu_i}\right)^{n_i} = c_N \, 4^{n_1} \, 3^{n_2} \, 2^{n_3}$$

if $n_1 + n_2 + n_3 = N$ and 0 otherwise.

At this point we have to pick c_N to make the sum of the probabilities equal to 1. This is not as easy as it sounds. To illustrate the problems involved consider the simple-sounding case $N = 4$. Our task in this case is to enumerate the triples (n_1, n_2, n_3) with $n_i \geq 0$ and $n_1 + n_2 + n_3 = 4$, compute $4^{n_1} 3^{n_2} 2^{n_3}$ and then sum up the weights to determine the normalizing constant.

4,0,0	256	2,0,2	64	0,4,0	81
3,1,0	192	1,3,0	108	0,3,1	54
3,0,1	128	1,2,1	72	0,2,2	36
2,2,0	144	1,1,2	48	0,1,3	24
2,1,1	96	1,0,3	32	0,0,4	16

Summing up the weights we get 1351, so multiplying the table above by $c_4 = 1/1351$ gives the stationary distribution.

One can get a somewhat better idea of the nature of the stationary distribution by looking at the distribution of the lengths of the individual queues. To simplify the table we just give the numerator of each probability. The denominator is always 1351. With this approach the numbers in the new table are just sums of ones in the old. For example, the probability of 3 customers in the second queue is sum of the probabilities for 1,3,0 and 0,3,1 or $108 + 54$.

Section 4.7 Closed Queueing Networks

	4	3	2	1	0
queue 1	256	320	304	260	211
queue 2	81	162	252	360	496
queue 3	16	56	148	350	781

Note that the second and third queues are quite often empty while queue 1 holds most of the palettes.

For $N = 10$ this gets to be considerably more complicated. To count the number of possible states, we observe that the number of (n_1, n_2, n_3) with integers $n_i \geq 0$ is the number of ways of arranging 10 o's and 2 x's in a row. To make the correspondence let n_1 be the number of o's before the first x, let n_2 be the number of o's between the first and the second x, and n_3 the number of o's after the second x. For example

$$o \ o \ o \ o \ o \ x \ o \ o \ x \ o \ o \ o$$

becomes $n_1 = 5$, $n_2 = 2$, and $n_3 = 3$. Having made this transformation it is clear that the number of possible states is the number of ways of picking 2 locations to put the x's or

$$\binom{12}{2} = \frac{12 \cdot 11}{2} = 66 \text{ states}$$

For each of these states we have to add up weight between $2^{10} = 1{,}024$ and $4^{10} = 1{,}048{,}576$, so the answer will be quite large.

Example 7.5. Machine repair. Consider the special case of Example 7.1 in which the breakdown rate is $\mu_1 = 1$ trucks per week, the service rates are $\mu_2 = 2$, $\mu_3 = 4$, and the routing matrix is

$$p(i,j) = \begin{pmatrix} 0 & 1/4 & 3/4 \\ 1 & 0 & 0 \\ 1 & 0 & 0 \end{pmatrix}$$

That is, 1/4 of the breakdowns are engine repairs, while 3/4 are tire repairs. To find the stationary distribution, we set $\pi_1 = c$, then compute

$$\pi_2 = (1/4)c \quad \pi_3 = (3/4)c$$

The sum of the π's is $2c$ so $c = 1/2$.

The first queue is an $M/M/\infty$ queue with service rate 1 per customer so

$$\psi_1(n) = \prod_{m=1}^{n} \varphi_i(m) = n!$$

The second and third queues are single servers with rates 2 and 4, so

$$\psi_2(n) = 2^n \quad \psi_3(n) = 4^n$$

Plugging into the formula in (7.2) we have that if $n_1 + n_2 + n_3 = N$

$$\pi(n_1, n_2, n_3) = c_N \frac{(1/2)^{n_1}}{n_1!} \frac{(1/8)^{n_2}}{2^{n_2}} \frac{(3/8)^{n_1}}{4^{n_3}}$$

$$= c'_N \cdot \frac{N!}{n_1!} 16^{n_1} 2^{n_2} 3^{n_3}$$

where in the last step we have multiplied the probabilities by $N! \, 32^N$ and changed the normalizing constant.

At this point we have to pick c'_N to make the sum of the probabilities equal to 1. To reduce the complexity from that of Example 7.4 we take $N = 3$. Enumerating the triples (n_1, n_2, n_3) with $n_i \geq 0$ and $n_1 + n_2 + n_3 = 3$ and then computing $16^{n_1} 2^{n_2} 3^{n_3} / n_1!$ gives the following result:

3,0,0	4096	1,2,0	384	0,3,0	48
2,1,0	1536	1,1,1	576	0,2,1	72
2,0,1	2304	1,0,2	864	0,1,2	128
				0,0,3	162

Summing up the weights we get 10,170, so multiplying the table above by $1/10,170$ gives the stationary distribution. From this we see that the number of broken trucks is 0, 1, 2, 3 with probabilities .403, .378, .179, .040.

4.8. Exercises

8.1. Imagine four individuals in a household, some of whom are healthy and some of whom are infected with "the flu." Imagine that infected individuals get well at rate μ, while each pair of persons has encounters at rate λ and the disease is transmitted if one is infected and the other susceptible. (a) Formulate a Markov chain model for this system. (b) Suppose to simplify things that recovery is impossible, i.e., $\mu = 0$. How long does it take from the time the first person is infected until all the members of the household are?

8.2. Solve the previous problem for a population of N individuals.

8.3. Consider a pure death process in which $i \to i - 1$ at rate μ when $i \geq 1$. Find the transition probability $p_t(i, j)$.

8.4. A salesman flies around between Atlanta, Boston, and Chicago as follows. She stays in each city for an exponential amount of time with mean 1/4 month if the city is A or B, but with mean 1/5 month if the city is C. From A she goes to B or C with probability 1/2 each; from B she goes to A with probability 3/4 and to C with probability 1/4; from C she always goes back to A. (a) Find the limiting fraction of time she spends in each city. (b) What is her average number of trips each year from Boston to Atlanta?

8.5. A small computer store has room to display up to 3 computers for sale. Customers come at times of a Poisson process with rate 2 per week to buy a computer and will buy one if at least 1 is available. When the store has only 1 computer left it places an order for 2 more computers. The order takes an exponentially distributed amount of time with mean 1 week to arrive. Of course, while the store is waiting for delivery, sales may reduce the inventory to 1 and then to 0. (a) Write down the matrix of transition rates Q_{ij} and solve $\pi Q = 0$ to find the stationary distribution. (b) At what rate does the store make sales?

8.6. Consider two machines that are maintained by a single repairman. Machine i functions for an exponentially distributed amount of time with rate λ_i before it fails. The repair times for each unit are exponential with rate μ_i. They are repaired in the order in which they fail. (a) Let X_t be the number of working machines at time t. Is X_t a Markov chain? (b) Formulate a Markov chain model for this situation with state space $\{0, 1, 2, 12, 21\}$. (c) Suppose that $\lambda_1 = 1$, $\mu_1 = 2$, $\lambda_2 = 3$, $\mu_2 = 4$. Find the stationary distribution.

8.7. Consider the set-up of the previous problem but now suppose machine 1 is much more important than 2, so the repairman will always service 1 if it is broken. (a) Formulate a Markov chain model for the this system with state space $\{0, 1, 2, 12\}$ where the numbers indicate the machines that are broken at the time. (b) Suppose that $\lambda_1 = 1$, $\mu_1 = 2$, $\lambda_2 = 3$, $\mu_2 = 4$. Find the stationary distribution.

8.8. Two people are working in a small office selling shares in a mutual fund. Each is either on the phone or not. Suppose that salesman i is on the phone for an exponential amount of time with rate μ_i and then off the phone for an exponential amount of time with rate λ_i. (a) Formulate a Markov chain model for this system with state space $\{0, 1, 2, 12\}$ where the state indicates who is on the phone. (b) Find the stationary distribution.

8.9. (a) Consider the special case of the previous problem in which $\lambda_1 = \lambda_2 = 1$, and $\mu_1 = \mu_2 = 3$, and find the stationary probabilities. (b) Suppose they upgrade their telephone system so that a call to one line that is busy is forwarded to the other phone and lost if that phone is busy. Find the new stationary probabilities.

8.10. Two people who prepare tax forms are working in a store at a local mall. Each has a chair next to his desk where customers can sit and be served. In addition there is one chair where customers can sit and wait. Customers arrive at rate λ but will go away if there is already someone sitting in the chair waiting. Suppose that server i requires an exponential amount of time with rate μ_i and that when both servers are free an arriving customer is equally likely to choose either one. (a) Formulate a Markov chain model for this system with state space $\{0, 1, 2, 12, 3\}$ where the first four states indicate the servers that are busy while the last indicates that there is a total of three customers in the system: one at each server and one waiting. (b) Consider the special case in which $\lambda = 2$, $\mu_1 = 3$ and $\mu_2 = 3$. Find the stationary distribution.

8.11. *Two queues in series.* Consider a two station queueing network in which arrivals only occur at the first server and do so at rate 2. If a customer finds server 1 free he enters the system; otherwise he goes away. When a customer is done at the first server he moves on to the second server if it is free and leaves the system if it is not. Suppose that server 1 serves at rate 4 while server 2 serves at rate 2. Formulate a Markov chain model for this system with state space $\{0, 1, 2, 12\}$ where the state indicates the servers who are busy. In the long run (a) what proportion of customers enter the system? (b) What proportion of the customers visit server 2?

8.12. A hemoglobin molecule can carry one oxygen or one carbon monoxide molecule. Suppose that the two types of gases arrive at rates λ_+ and λ_- and attach for an exponential amount of time with rates μ_+ and μ_-, respectively. Formulate a Markov chain model with state space $\{+, 0, -\}$ where + denotes an attached oxygen molecule, − an attached carbon monoxide molecule, and 0 a free hemoglobin molecule and find the long-run fraction of time the hemoglobin molecule is in each of its three states.

8.13. A machine is subject to failures of types $i = 1, 2, 3$ at rates λ_i and a failure of type i takes an exponential amount of time with rate μ_i to repair. Formulate a Markov chain model with state space $\{0, 1, 2, 3\}$ and find its stationary distribution.

8.14. Solve the previous problem in the concrete case $\lambda_1 = 1/24$, $\lambda_2 = 1/30$, $\lambda_3 = 1/84$, $\mu_1 = 1/3$, $\mu_2 = 1/5$, and $\mu_3 = 1/7$.

8.15. Customers arrive at a full-service one-pump gas station at rate of 20 cars per hour. However, customers will go to another station if there are at least two cars in the station, i.e., one being served and one waiting. Suppose that the service time for customers is exponential with mean 6 minutes. (a) Formulate a Markov chain model for the number of cars at the gas station and find its stationary distribution. (b) On the average how many customers are served per hour?

8.16. Solve the previous problem for a two-pump self-serve station under the assumption that customers will go to another station if there are at least four cars in the station, i.e., two being served and two waiting.

8.17. Three frogs are playing near a pond. When they are in the sun they get too hot and jump in the lake at rate 1. When they are in the lake they get too cold and jump onto the land at rate 2. Let X_t be the number of frogs in the sun at time t. (a) Find the stationary distribution for X_t. (b) Check the answer to (a) by noting that the three frogs are independent two-state Markov chains.

8.18. A computer lab has three laser printers, two that are hooked to the network and one that is used as a spare. A working printer will function for an exponential amount of time with mean 20 days. Upon failure it is immediately sent to the repair facility and replaced by another machine if there is one in working order. At the repair facility machines are worked on by a single repairman who needs an exponentially distributed amount of time with mean 2 days to fix one printer. In the long run how often are there two working printers?

8.19. A computer lab has three laser printers that are hooked to the network. A working printer will function for an exponential amount of time with mean 20 days. Upon failure it is immediately sent to the repair facility. There machines are worked on by two repairman who can each repair one printer in an exponential amount of time with mean 2 days. However, it is not possible for two people to work on one printer at once. (a) Formulate a Markov chain model for the number of working printers and find the stationary distribution. (b) How often are both repairmen busy? (c) What is the average number of machines in use?

8.20. Consider a barbershop with two barbers and two waiting chairs. Customers arrive at a rate of 5 per hour. Customers arriving to a fully occupied shop leave without being served. Find the stationary distribution for the number of customers in the shop, assuming that the service rate for each barber is 2 customers per hour.

8.21. Consider a barbershop with one barber who can cut hair at rate 4 and three waiting chairs. Customers arrive at a rate of 5 per hour. (a) Argue that this new set-up will result in fewer lost customers than the previous scheme. (b) Compute the increase in the number of customers served per hour.

8.22. There are two tennis courts. Pairs of players arrive at rate 3 per hour and play for an exponentially distributed amount of time with mean 1 hour. If there are already two pairs of players waiting new arrivals will leave. Find the stationary distribution for the number of courts occupied.

8.23. A taxi company has three cabs. Calls come in to the dispatcher at times of a Poisson process with rate 2 per hour. Suppose that each requires an exponential amount of time with mean 20 minutes, and that callers will hang up if they hear there are no cabs available. (a) What is the probability all three cabs are busy when a call comes in? (b) In the long run, on the average how many customers are served per hour?

8.24. Consider a taxi station at an airport where taxis and (groups of) customers arrive at times of Poisson processes with rates 2 and 3 per minute. Suppose that a taxi will wait no matter how many other taxis are present. However, if an arriving person does not find a taxi waiting he leaves to find alternative transportation. (a) Find the proportion of arriving customers that get taxis. (b) Find the average number of taxis waiting.

8.25. *Queue with impatient customers.* Customers arrive at a single server at rate λ and require an exponential amount of service with rate μ. Customers waiting in line are impatient and if they are not in service they will leave at rate δ independent of their position in the queue. (a) Show that for any $\delta > 0$ the system has a stationary distribution. (b) Find the stationary distribution in the very special case in which $\delta = \mu$.

8.26. Customers arrive at the Shortstop convenience store at a rate of 20 per hour. When two or fewer customers are present in the checkout line, a single clerk works and the service time is 3 minutes. However, when there are three or more customers are present, an assistant comes over to bag up the groceries and reduces the service time to 2 minutes. Assuming the service times are exponentially distributed, find the stationary distribution.

8.27. Customers arrive at a carnival ride at rate λ. The ride takes an exponential amount of time with rate μ, but when it is in use, the ride is subject to breakdowns at rate α. When a breakdown occurs all of the people leave since they know that the time to fix a breakdown is exponentially distributed with rate β. (i) Formulate a Markov chain model with state space $\{-1, 0, 1, 2, \ldots\}$ where -1 is broken and the states $0, 1, 2, \ldots$ indicate the number of people waiting or in service. (ii) Show that the chain has a stationary distribution of the form $\pi(-1) = a$, $\pi(n) = b\theta^n$ for $n \geq 0$.

8.28. *Logistic model.* Consider the birth and death chain with birth rates $\lambda_n = \alpha + \beta n$ and death rates $\mu_n = (\delta + \gamma n)n$ where all four constants are positive. In words, there is immigration at rate α, each particle gives birth at rate β and dies at rate $\delta + \gamma n$, i.e., at a rate which starts at rate δ and increases linearly due to crowding. Show that the system has a stationary distribution.

8.29. Customers arrive at a two-server station according to a Poisson process with rate λ. Upon arriving they join a single queue to wait for the next available server. Suppose that the service times of the two servers are exponential with rates μ_a and μ_b and that a customer who arrives to find the system empty will go to each of the servers with probability 1/2. Formulate a Markov chain model for this system with state space $\{0, a, b, 2, 3, \ldots\}$ where the states give the number of customers in the system, with a or b indicating there is one customer at a or b respectively. Show that this system is time reversible. Set $\pi(2) = c$ and solve to find the limiting probabilities in terms of c.

8.30. *Detailed balance for three state chains.* Consider a chain with state space $\{1, 2, 3\}$ in which $q(i, j) > 0$ if $i \neq j$ and suppose that there is a stationary distribution that satisfies the detailed balance condition. (a) Let $\pi(1) = c$. Use the detailed balance condition between 1 and 2 to find $\pi(2)$ and between 2 and 3 to find $\pi(3)$. (b) What conditions on the rates must be satisfied for there to be detailed balance between 1 and 3?

8.31. *Kolmogorov cycle condition.* Consider an irreducible Markov chain with state space S. We say that the cycle condition is satisfied if given a cycle of states $x_0, x_1, \ldots, x_n = x_0$ with $q(x_{i-1}, x_i) > 0$ for $1 \leq i \leq n$, we have

$$\prod_{i=1}^{n} q(x_{i-1}, x_i) = \prod_{i=1}^{n} q(x_i, x_{i-1})$$

(a) Show that if q has a stationary distribution that satisfies the detailed balance condition, then the cycle condition holds. (b) To prove the converse, suppose that the cycle condition holds. Let $a \in S$ and set $\pi(a) = c$. For $b \neq a$ in S let $x_0 = a, x_1 \ldots x_k = b$ be a path from a to b with $q(x_{i-1}, x_i) > 0$ for $1 \leq i \leq k$ let

$$\pi(b) = \prod_{j=1}^{k} \frac{q(x_{i-1}, x_i)}{q(x_i, x_{i-1})}$$

Show that $\pi(b)$ is well defined, i.e., is independent of the path chosen. Then conclude that π satisfies the detailed balance condition.

8.32. Let X_t be a Markov chain with a stationary distribution π that satisfies the detailed balance condition. Let Y_t be the chain constrained to stay in a subset A of the state space. That is, jumps which take the chain out of A are not allowed, but allowed jumps occur at the original rates. In symbols, $\bar{q}(x, y) = q(x, y)$ if $x, y \in A$ and 0 otherwise. Let $C = \sum_{y \in A} \pi(y)$. Show that $\nu(x) = \pi(x)/C$ is a stationary distribution for Y_t.

8.33. Two barbers share a common waiting room that has N chairs. Barber i gives service at rate μ_i and has customers that arrive at rate $\lambda_i < \mu_i$. Assume that customers always wait for the barber they came to see even when the other is free, but go away if the waiting room is full. Let N_t^i be the number of customers for barber i that are waiting or being served. Find the stationary distribution for (N_t^1, N_t^2).

8.34. Solve the previous problem when $\lambda_1 = 1$, $\mu_1 = 3$, $\lambda_2 = 2$, $\mu_2 = 4$, and $N = 2$.

8.35. Consider an $M/M/s$ queue with no waiting room. In words, requests for a phone line occur at a rate λ. If one of the s lines is free, the customer takes it and talks for an exponential amount of time with rate μ. If no lines are free, the customer goes away never to come back. Find the stationary distribution. You do not have to evaluate the normalizing constant.

8.36. There are 15 lily pads and 6 frogs. Each frog at rate 1 gets the urge to jump and when it does, it moves to one of the 9 vacant pads chosen at random. Find the stationary distribution for the set of occupied lily pads.

8.37. *Simple exclusion process.* There are 1278 particles on a 100×100 checkerboard. Each at rate one attempts to jump, and when it does it executes a random movement in one of the four directions up, down, left, and right. If the proposed move would take the particle off the board or onto a square occupied by another particle the jump is suppressed and nothing happens (hence the term exclusion). Find the stationary distribution for the set of positions occupied by particles.

8.38. Consider a production system consisting of a machine center followed by an inspection station. Arrivals from outside the system occur only at the machine center and follow a Poisson process with rate λ. The machine center and inspection station are each single-server operations with rates μ_1 and μ_2. Suppose that each item independently passes inspection with probability p. When an object fails inspection it is sent to the machine center for reworking. Find the conditions on the parameters that are necessary for the system to have a stationary distribution.

8.39. Consider a three station queueing network in which arrivals to servers $i = 1, 2, 3$ occur at rates $3, 2, 1$, while service at stations $i = 1, 2, 3$ occurs at rates $4, 5, 6$. Suppose that the probability of going to j when exiting i, $p(i, j)$ is given by $p(1, 2) = 1/3$, $p(1, 3) = 1/3$, $p(2, 3) = 2/3$, and $p(i, j) = 0$ otherwise. Find the stationary distribution.

8.40. *Feed-forward queues.* Consider a k station queueing network in which arrivals to server i occur at rate λ_i and service at station i occurs at rate μ_i. We say that the queueing network is feed-forward if the probability of going from i to $j < i$ has $p(i, j) = 0$. Consider a general three station feed-forward queue. What conditions on the rates must be satisfied for a stationary distribution to exist?

8.41. *Queues in series.* Consider a k station queueing network in which arrivals to server i occur at rate λ_i and service at station i occurs at rate μ_i. In this problem we examine the special case of the feed-forward system in which

$p(i, i+1) = p_i$ for $1 \leq i < k$. In words the customer goes to the next station or leaves the system. What conditions on the rates must be satisfied for a stationary distribution to exist?

8.42. At registration at a very small college, students arrive at the English table at rate 10 and at the Math table at rate 5. A student who completes service at the English table goes to the Math table with probability 1/4 and to the cashier with probability 3/4. A student who completes service at the Math table goes to the English table with probability 2/5 and to the cashier with probability 3/5. Students who reach the cashier leave the system after they pay. Suppose that the service times for the English table, Math table, and cashier are 25, 30, and 20, respectively. Find the stationary distribution.

8.43. Three vendors have vegetable stands in a row. Customers arrive at the stands 1, 2, and 3 at rates 10, 8, and 6. A customer visiting stand 1 buys something and leaves with probability 1/2 or visits stand 2 with probability 1/2. A customer visiting stand 3 buys something and leaves with probability 7/10 or visits stand 2 with probability 3/10. A customer visiting stand 2 buys something and leaves with probability 4/10 or visits stands 1 or 3 with probability 3/10 each. Suppose that the service rates at the three stands are large enough so that a stationary distribution exists. At what rate do the three stands make sales. To check your answer note that since each entering customers buys exactly once the three rates must add up to 10+8+6=24.

8.44. Four children are playing two video games. The first game, which takes an average of 4 minutes to play, is not very exciting, so when a child completes a turn on it they always stand in line to play the other one. The second one, which takes an average of 8 minutes, is more interesting so when they are done they will get back in line to play it with probability 1/2 or go to the other machine with probability 1/2. Assuming that the turns take an exponentially distributed amount of time, find the stationary distribution of the number of children playing or in line at each of the two machines.

8.45. A computer lab has 3 laser printers and 5 toner cartiridges. Each machine requires one toner cartridges which lasts for an exponentially distributed amount of time with mean 6 days. When a toner cartridge is empty it is sent to a repairman who takes an exponential amount of time with mean 1 day to refill it. This system can be modeled as a closed queueing network with 5 customers (the toner cartridges), one $M/M/3$ queue, and one $M/M/1$ queue. Use this observation to compute the stationary distribution. How often are all three printers working?

8.46. The exercise room at the Wonderland motel has three pieces of equipment. Five businessmen who are trapped there by one of Ithaca's snowstorms use machines 1,2,3 for an exponential amount of time with means 15,10,5 minutes.

When a person is done with one piece of equipment, he picks one of the other two at random. If it is occupied he stands in line to use it. Let (n_1, n_2, n_3) be the number of people using or in line to use each of the three pieces of equipment. (a) Find the stationary distribution. (b) Evaluate the norming constant.

8.47. Generalize the previous problem so that there are N machines and M businessmean, but simplify it by supposing that all machines have the same service time. Find the stationary distribution.

5 Renewal Theory

5.1. Basic Definitions

In the Poisson process the times between successive arrivals are independent and exponentially distributed. The lack of memory property of the exponential distribution is crucial for many of the special properties of the Poisson process derived in Chapter 3. However, in many situations the assumption of exponential interarrival times is not justified. In this section we will consider a generalization of Poisson processes called **renewal processes** in which the times $t_1, t_2, \ldots$ between events are independent and have distribution F.

In order to have a simple metaphor with which to discuss renewal processes, we will think of a single light bulb maintained by a very diligent janitor, who replaces the light bulb immediately after it burns out. Let t_i be the lifetime of the ith light bulb. We assume that the light bulbs are bought from one manufacturer, so we suppose

$$P(t_i \leq t) = F(t)$$

where F is a distribution function with $F(0) = P(t_i \leq 0) = 0$.

If we start with a new bulb (numbered 1) at time 0 and each light bulb is replaced when it burns out, then $T_n = t_1 + \cdots + t_n$ gives the time that the nth bulb burns out, and

$$N(t) = \max\{n : T_n \leq t\}$$

is the number of light bulbs that have been replaced by time t.

If renewal theory were only about changing light bulbs, it would not be a very useful subject. The reason for our interest in this system is that it captures the essence of a number of different situations. The first and most obvious generalization of the light bulb story is

Example 1.1. Machine repair. Instead of a light bulb, think of a machine that works for an amount of time s_i before it fails, requiring an amount of time

u_i to be repaired. Let $t_i = s_i + u_i$ be the length of the ith cycle of breakdown and repair. If we assume that the repair leaves the machine in a "like new" condition, then the t_i are independent and identically distributed (i.i.d.) and a renewal process results.

Example 1.2. Traffic flow. The distances between cars on one lane of a two (or more) lane road are often assumed to form a renewal processes. So are the time durations between consecutive cars passing a fixed location.

Example 1.3. M/M/1 queue. Suppose that a queue starts empty at time 0. Let T_n be the nth time the queue becomes empty. That is, the nth moment in time when the queue had one customer who then departed. The strong Markov property implies that the times $t_n = T_n - T_{n-1}$ are independent, so T_n is a renewal process.

Example 1.4. Markov chains. Generalizing the previous example, let X_t be a continuous-time Markov chain and suppose that $X_0 = i$. Let T_n be the nth time that the process returns to i. Again, the strong Markov property implies that $t_n = T_n - T_{n-1}$ are independent, so T_n is a renewal process.

Example 1.5. Counter processes. The following situation arises, for example, in medical imaging applications. Particles arrive at a counter at times of a Poisson process with rate λ. Each arriving particle that finds the counter free gets registered and locks the counter for an amount of time τ. Particles arriving during the locked period have no effect. If we assume the counter starts in the unlocked state, then the times T_n at which it becomes unlocked for the nth time form a renewal process.

Abstracting from the examples listed above, T_n is called the time of the nth **renewal**. We say that the renewal process "starts afresh" at time T_n, since the process from T_n onward looks like a copy of the original process. $N(t)$ is called the number of renewals by time t. In the picture $N(t) = 4$.

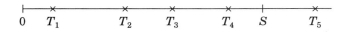

Renewal processes can be described by either giving the arrival times T_n or the counting process $N(t)$. The quantities are related by

(1.1) $$\{N(t) \geq n\} = \{T_n \leq t\}$$

Proof. There are at least n renewals by time t if and only if the nth renewal occurs before time t, meaning $\leq t$. □

Using (1.1) we can compute the expected number of renewals up to time t. The first step is the following result.

(1.2) **Lemma.** If $X \geq 0$ then $EX = \int_0^\infty P(X > t)\, dt$.

Why is this true? Let $i_X(t) = 1$ if $0 < t < X$, 0 otherwise. From the definition it is easy to see that $\int_0^\infty i_X(t)\, dt = \int_0^X 1\, dt = X$. Taking the expected value of each side, putting the expected value inside the integral, and noting that $Ei_X(t) = P(X > t)$, we have

$$EX = \int_0^\infty Ei_X(t)\, dt = \int_0^\infty P(X > t)\, dt \quad \square$$

An important special case of (1.2) is:

(1.3) **Lemma.** If X is a nonnegative integer valued random variable, then

$$EX = \sum_{n=1}^\infty P(X \geq n)$$

Proof. If $n - 1 \leq t < n$, then $\{X > t\} = \{X \geq n\}$. This means that the function $t \to P(X > t)$ is $P(X \geq 1)$ for $0 \leq t < 1$, $P(X \geq 2)$ for $1 \leq t < 2$, etc. So

$$\int_0^\infty P(X > t)\, dt = \sum_{n=1}^\infty P(X \geq n) \quad \square$$

Using (1.3) with (1.1), it follows that

(1.4) $$EN(t) = \sum_{n=1}^\infty P(N(t) \geq n) = \sum_{n=1}^\infty P(T_n \leq t)$$

If the t_i are exponential with rate λ, then T_n has a gamma(n, λ) distribution and $EN(t) = \lambda t$. However, for most distributions F, we do not have good formulas for T_n and cannot find explicit formulas for $EN(t)$. A few of the exceptions to this rule can be found in the problems at the end of the chapter.

We can get some information about $EN(t)$ by using Wald's equation, (4.3) of Chapter 2. For the readers who might have skipped that chapter, we now recall the necessary definitions and give another proof of that result. Let $S_n = x_1 + \cdots + x_n$, where $x_1, x_2, \ldots x_n$ are i.i.d. with $\mu = Ex_i$. We say N is a **stopping time**, if the occurrence (or nonoccurrence) of the event "we stop at time n" can be determined by looking at the values of $S_1, S_2 \ldots, S_n$.

The basic linearity property of expected value implies that $ES_n = n\mu$, i.e., the mean of the sum of n random variables is n times the mean of one term. The next result says that this result also holds for stopping times that have finite mean.

(1.5) Wald's equation. *If N is a stopping time with $EN < \infty$, then*

$$ES_N = \mu EN$$

Proof of (1.5). Let $i_N(m) = 1$ if $N \geq m$ and 0 otherwise. From the definition it is easy to see that

$$S_N = \sum_{m=1}^{\infty} x_m i_N(m)$$

Now $\{N \geq m\}^c = \cup_{k=1}^{m-1}\{N = k\}$, so from the definition of stopping time it follows that this event this can be determined from $x_1, \ldots x_{m-1}$, and hence is independent of x_m. Taking expected values now, we have

$$ES_N = \sum_{m=1}^{\infty} Ex_m \cdot Ei_N(m) = \mu \sum_{m=1}^{\infty} P(N \geq m) = \mu EN$$

where in the last two steps we have used the facts that $Ex_m = \mu$ and $Ei_N(m) = P(N \geq m)$, then the formula in (1.3). □

Bounds on $EN(t)$. For our first application of (1.5), we note that

$$N = \min\{n : T_n > t\} = N(t) + 1$$

is a stopping time. In words, we can stop the process the first time T_n exceeds t. In contrast, note that $N(t) = \max\{n : T_n \leq t\}$ is not a stopping time since

$$\{N(t) = n\} = \{T_n \leq t < T_{n+1}\}$$

requires knowledge of t_{n+1}. Using (1.5) with $N = N(t) + 1$ we have

(1.6) $$ET_{N(t)+1} = \mu E(N(t) + 1)$$

Since $T_{N(t)+1} > t$, it follows from the last result that

(1.7) $$EN(t) + 1 \geq \frac{t}{\mu}$$

To get a result in the other direction we need a bound on the difference $T_{N(t)+1} - t$. The simplest approach is

(1.8) **Lemma.** *If $t_i \leq M$, then*

$$E(N(t) + 1) \leq (t + M)/\mu$$

Proof. The assumption implies $T_{N(t)+1} \leq t + M$, so the desired result follows from (1.6). □

To illustrate the use of (1.7) and (1.8) consider

Example 1.6. Bill has a portable CD player that works on a single battery. If the lifetime of a battery is uniformly distributed over 30 to 60 days, then on the average how many batteries will he use up in 10 years? (Suppose for simplicity a year has 360 days.)

Solution. The first step is to note that the mean life of a battery is 45 days. One can see this by noting that $45 = (30+60)/2$ is the midpoint of the interval, so symmetry dictates it must be the mean. To calculate this directly from the definition, we start with the observation that the density function for the life of the ith battery, t_i, is

$$f(x) = \begin{cases} 1/30 & \text{for } x \in [30, 60] \\ 0 & \text{otherwise} \end{cases}$$

So integrating gives

$$Et_i = \int_{30}^{60} x \cdot \frac{1}{30} \, dx = \frac{x^2}{60}\Big|_{30}^{60} = \frac{60^2}{60} - \frac{30^2}{60} = \frac{2700}{60} = 45$$

Using the bounds in (1.6) we get

$$EN(3600) + 1 \geq 3600/45 = 80$$
$$EN(3600) + 1 \leq 3660/45 = 81.333$$

so $79 \leq EN(3600) \leq 80.333$. □

5.2. Laws of Large Numbers

To avoid the difficulty that exact formulas for $EN(t)$ are very complicated, we turn our attention to studying the asymptotic behavior of the number of renewals, $N(t)$, as time tends to infinity. Our first result is a strong law of large numbers:

(2.1) Theorem. Let $\mu = Et_i$ be mean interarrival time. With probability one,

$$N(t)/t \to 1/\mu \quad \text{as } t \to \infty$$

In words, this says that if our light bulb lasts μ years on the average then in t years we will use up about t/μ light bulbs. A special case of (2.1) is:

Example 2.1. Strong law for the Poisson process. If the interarrival times t_i are independent exponentials with rate λ, then $Et_i = 1/\lambda$. In this case (2.1) implies that if $N(t)$ is the number of arrivals up to time t in a Poisson process, then

$$N(t)/t \to \lambda \quad \text{as } t \to \infty$$

Proof of (2.1). We use the

(2.2) Strong law of large numbers. Let $x_1, x_2, x_3, \ldots$ be i.i.d. with $Ex_i = \mu$, and let $S_n = x_1 + \cdots + x_n$. Then with probability one,

$$S_n/n \to \mu \quad \text{as } n \to \infty$$

Taking $x_i = t_i$, we have $S_n = T_n$, so (2.2) implies that with probability one,

$$T_n/n \to \mu \quad \text{as } n \to \infty$$

Now by definition,

$$T_{N(t)} \leq t < T_{N(t)+1}$$

Dividing by $N(t)$, we have

$$\frac{T_{N(t)}}{N(t)} \leq \frac{t}{N(t)} \leq \frac{T_{N(t)+1}}{N(t)+1} \cdot \frac{N(t)+1}{N(t)}$$

By the strong law of large numbers, the left- and right-hand sides converge to μ. From this it follows that $t/N(t) \to \mu$ and hence $N(t)/t \to 1/\mu$. □

(2.1) says that $N(t)/t \to 1/\mu$. Taking expected value and not worrying about the details, we see that

$$EN(t)/t \to 1/\mu$$

The next results says that this reasoning is correct.

(2.3) Theorem. *Let $\mu = Et_i$ be mean interarrival time, and $m(t) = EN(t)$ be the expected number of renewals by time t. Then*

$$m(t)/t \to 1/\mu \quad \text{as } t \to \infty$$

In a sense (2.3) has a weaker conclusion than (2.1), since it concerns the expected value of $N(t)$ rather than the random variables themselves. However, as the next example illustrates, we may have $Y_t \to c$ without $EY_t \to c$.

Example 2.2. Let U be uniform on $(0,1)$ and let

$$Y_t = \begin{cases} c+t & \text{if } U \leq 1/t \\ c & \text{if } U > 1/t \end{cases}$$

Since $U > 0$ with probability one, we have $Y_t \to c$ as $t \to \infty$ but for $t \geq 1$

$$EY_t = cP(U > 1/t) + (c+t)P(U \leq 1/t) = c + t \cdot \frac{1}{t} = c+1$$

and so $EY_t \to c+1 > c$. □

This example shows that we may have $Y_t \to c$ without $EY_t \to c$. The problem should be obvious: a significant amount of the expected value comes from an event of small probability. It may seen paranoid to think that this sort of pathology will occur in the distribution of N_t/t, but to establish (2.3) we have to show that it does not happen. To do this we rely on the bounds established in Section 5.1.

Proof of (2.3). (1.7) implies that

$$EN(t) + 1 \geq \frac{t}{\mu}$$

Dividing by t, letting $t \to \infty$, and using the fact that $1/t \to 0$, we have

($\star$) $$\lim_{t \to \infty} \frac{EN(t)}{t} \geq \frac{1}{\mu}$$

To get a result in the other direction requires more work. Let $\bar{t}_i = \min\{t_i, M\}$, let $\mu_M = E\bar{t}_i$ and let $\bar{T}_n = \bar{t}_1 + \cdots + \bar{t}_n$. Since $\bar{t}_i \leq t_i$, we have $\bar{T}_n \leq T_n$ and $\bar{N}(t) \geq N(t)$. Since $\bar{t}_i \leq M$, it follows from (1.8) that

$$EN(t) \leq E\bar{N}(t) + 1 \leq \frac{t+M}{\mu_M}$$

Dividing by t and letting $t \to \infty$, we have

$$\lim_{t\to\infty} \frac{EN(t)}{t} \leq \frac{1}{\mu_M}$$

Letting $M \uparrow \infty$ and noting that $\mu_M = E(t_i \wedge M) \uparrow Et_i = \mu$, we have

($\star\star$)
$$\lim_{t\to\infty} \frac{EN(t)}{t} \leq \frac{1}{\mu}$$

which with ($\star$) gives the desired result. $\square$

Our next topic is a simple extension of the notion of a renewal process that greatly extends the class of possible applications. We suppose that at the time of the ith renewal we earn a reward r_i. The reward r_i may depend on the ith interarrival time t_i, but we will assume that the pairs (r_i, t_i), $i = 1, 2, \ldots$ are independent and have the same distribution. Let

$$R(t) = \sum_{i=1}^{N(t)} r_i$$

be the total amount of rewards earned by time t. The main result about renewal reward processes is the following strong law of large numbers.

(2.4) Theorem. *With probability one,*

$$\frac{R(t)}{t} \to \frac{Er_i}{Et_i}$$

Proof. Multiplying and dividing by $N(t)$, we have

$$\frac{R(t)}{t} = \left(\frac{1}{N(t)} \sum_{i=1}^{N(t)} r_i\right) \frac{N(t)}{t} \to Er_i \cdot \frac{1}{Et_i}$$

where in the last step we have used (2.1) and applied the strong law of large numbers to the sequence r_i. □

Why is (2.4) reasonable? If we say that a cycle is completed every time a renewal occurs, then the limiting rate at which we earn rewards is

$$\frac{\text{expected reward per cycle}}{\text{expected time per cycle}}$$

Even less formally, this can be written as

$$\text{reward/time} = \frac{\text{expected reward/cycle}}{\text{expected time/cycle}}$$

an equation that can be "proved" by pretending the words on the right-hand side are numbers and then canceling the "expected" and "1/cycle" that appear in numerator and denominator. □

The last calculation is not given to convince you that (2.4) is correct but to help you remember the result. With that goal in mind we also introduce:

Example 2.3. Nonrandom times and rewards. If we earn a reward of ρ dollar every τ units of time then in the long run we earn ρ/τ dollars per unit time. To get from this to the answer given in (2.4), note that the answer there only depends on the means Er_i and Et_i, so the general answer must be

$$\rho/\tau = Er_i/Et_i$$

This device can be applied to remember many of the results in this chapter: when the answer only depends on the mean the limit must be the same as in the case when the times are not random.

To illustrate the use of (2.4) we consider

Example 2.4. Long run car costs. Suppose that the lifetime of a car is a random variable with density function h. Our methodical Mr. Brown buys a new car as soon as the old one breaks down or reaches T years. Suppose that a new car costs A dollars and that an additional cost of B dollars to repair the vehicle is incurred if it breaks down before time T. What is the long-run cost per unit time of Mr. Brown's policy?

Solution. The duration of the ith cycle, t_i, has

$$Et_i = \int_0^T th(t)\,dt + T\int_T^\infty h(t)\,dt$$

218 Chapter 5 Renewal Theory

since the length of the cycle will be t_i if the car's life is $t < T$, but T if the car's life $t \geq T$. The reward (or cost) of the ith cycle has

$$Er_i = A + B \int_0^T h(t)\,dt$$

since Mr. Brown always has to pay A dollars for a new car but only owes the additional B dollars if the car breaks down before time T. Using (2.1) we see that the long run cost per unit time is

$$\frac{Er_i}{Et_i} = \frac{A + B \int_0^T h(t)\,dt}{\int_0^T th(t)\,dt + \int_T^\infty Th(t)\,dt}$$

Concrete example. Suppose that the lifetime of Mr. Brown's car is exponentially distributed with mean 4 years, i.e., $h(t) = (1/4)e^{-t/4}$. We should pause here to acknowledge that this is not a reasonable assumption, since when cars get older they have a greater tendency to break. However, having confessed to this weakness, we will proceed with this assumption since it makes calculations easier, and we expect the results for the exponential to be typical of those for other distributions.

Suppose that the replacement time Mr. Brown chooses is $T = 5$ years, and suppose that the cost of a new car is $A = 10$ (thousand dollars), while the breakdown cost is $B = 1.5$ (thousand dollars). In this case the expected cost per cycle

$$Er_i = 10 + 1.5 \int_0^T \frac{1}{4} e^{-t/4}\,dt = 10 + 1.5 \cdot (1 - e^{-5/4})$$
$$= 10 + 1.5(.7135) = 11.070 \quad \text{thousand dollars}$$

while the expected cycle length is

$$Et_i = \int_0^5 t \cdot \frac{1}{4} e^{-t/4}\,dt + \int_5^\infty 5 \cdot \frac{1}{4} e^{-t/4}\,dt$$
$$= \int_0^\infty t \cdot \frac{1}{4} e^{-t/4}\,dt + \int_5^\infty (5-t) \cdot \frac{1}{4} e^{-t/4}\,dt$$

Here we have added and subtracted the integral of $(t/4)e^{-t/4}$ over $[5,\infty)$ to see that the first term is just the mean, 4, while the second is -1 times the amount of time the car lasts after 5 years. By the lack of memory property this must be $4P(t_i > 5) = 4e^{-5/4}$ and we have

$$Et_i = 4 - 4e^{-5/4} = 4(.7135) = 2.854 \quad \text{years.}$$

One can of course also arrive at this answer by doing calculus on the original expression for Et_i. We like the approach followed above since it shows that the cycle length is the original lifetime minus what we lose by trading in a car that is still working at time 5. Combining the expressions for the Er_i and Et_i we see that the long-run cost per unit time is

$$\frac{Er_i}{Et_i} = \frac{11.070}{2.854} = 3.879 \text{ thousand dollars per year.} \qquad \square$$

Using the idea of renewal reward processes, we can easily treat the following extension of renewal processes.

Example 2.5. Alternating renewal processes. Let $s_1, s_2, \ldots$ be independent with a distribution F that has mean μ_F, and let $u_1, u_2, \ldots$ be independent with distribution G that has mean μ_G. For a concrete example consider the machine in Example 1.1 that works for an amount of time s_i before needing a repair that takes u_i units of time. However, to talk about things in general we will say that the alternating renewal process spends an amount of time s_i in state 1, an amount of time u_i in state 2, and then repeats the cycle again.

(2.5) Theorem. *In an alternating renewal process, the limiting fraction of time in state 1 is*

$$\frac{\mu_F}{\mu_F + \mu_G}$$

To see that this is reasonable and to help remember the formula, consider the nonrandom Example 2.3. If the machine always works for exactly μ_F days and then needs repair for exactly μ_G days, then the limiting fraction of time spent working is $\mu_F/(\mu_F + \mu_G)$.

Why is this true? In order to compute the limiting fraction of time the machine is working we let $t_i = s_i + u_i$ be the duration of the ith cycle, and let the reward $r_i = s_i$, the amount of time the machine was working during the ith cycle. In this case, if we ignore the contribution from the cycle that is in progress at time t, then

$$R(t) = \sum_{i=1}^{N(t)} r_i$$

will be the amount of time the machine has been working up to time t. Thus, (2.4) implies that

$$\frac{R(t)}{t} \to \frac{Er_i}{Et_i} = \frac{\mu_F}{\mu_F + \mu_G} \qquad \square$$

The standard hand-waving. The phrase "if we ignore the contribution from the cycle that is in progress at time t" will occur so many times in this section that we have given it a bold-faced name. One can fill in the missing details in the argument for (2.5) by following the outline in the next exercise. However, the details here are already somewhat technical and will get worse as our story progresses, so we will ignore them in the future.

EXERCISE 2.1. Let $r_1, r_2, \ldots$ be independent nonnegative and have $Er_i < \infty$.
(i) Show that for any $\epsilon > 0$

$$\sum_{n=1}^{\infty} P(r_n > n\epsilon) \leq \int_0^{\infty} P(r_n/\epsilon > t)\, dt \leq E(r_i/\epsilon) < \infty$$

(ii) Use (i) to conclude that $r_n/n \to 0$ and hence that $\frac{1}{n} \max_{1 \leq m \leq n} r_m \to 0$.
(iii) Show that the result in (ii) implies $r_{N(t)}/t \to 0$.

For a concrete example of alternating renewal processes, consider

Example 2.6. Poisson janitor. A light bulb burns for an amount of time having distribution F with mean μ_F then burns out. A janitor comes at times of a rate λ Poisson process to check the bulb and will replace the bulb if it is burnt out. (a) At what rate are bulbs replaced? (c) What is the limiting fraction of time that the light bulb works? (b) What is the limiting fraction of visits on which the bulb is working?

Solution. Suppose that a new bulb is put in at time 0. It will last for an amount of time s_1. Using the lack of memory property of the exponential distribution, it follows that the amount of time until the next inspection, u_1, will have an exponential distribution with rate λ. The bulb is then replaced and the cycle starts again, so we have an alternating renewal process.

To answer (a), we note that the expected length of a cycle is $\mu_F + 1/\lambda$, so if $N(t)$ is the number of bulbs replaced by time t, then it follows from (1.2) that

$$\frac{N(t)}{t} \to \frac{1}{\mu_F + 1/\lambda}$$

In words, bulbs are replaced on the average every $\mu_F + 1/\lambda$ units of time.

To answer (b), we let $r_i = s_i$, so (2.5) implies that in the long run, the fraction of time the bulb has been working up to time t is

$$\frac{\mu_F}{\mu_F + 1/\lambda}$$

To answer (c), we note that if $V(t)$ is the number of visits the janitor has made by time t, then by the law of large numbers for the Poisson process given in Example 2.1 we have
$$\frac{V(t)}{t} \to \lambda$$
Combining this with the result of (a), we see that the fraction of visits on which bulbs are replaced
$$\frac{N(t)}{V(t)} \to \frac{1/(\mu_F + 1/\lambda)}{\lambda} = \frac{1/\lambda}{\mu_F + 1/\lambda}$$
This answer is reasonable since it is also the limiting fraction of time the bulb is off.

EXERCISE 2.2. To derive the answer to (c) another way, let I be the number of inspections that occur during the lifetime of one bulb. Show that $EI = \lambda \mu_F$. Taking into account the one visit at the beginning of the bulb's life, which is the only one that leads to a replacement, we see that the limiting fraction of inspections that lead to replacement should be $1/(1 + \lambda \mu_F)$.

5.3. Applications to Queueing Theory

In this section we will use the ideas of renewal theory to prove results for queueing systems with general service times. In the first part of this section we will consider general arrival times; in the second we will specialize to Poisson inputs.

a. GI/G/1 queue.

Here the GI stands for general input. That is, we suppose that the times between successive arrivals are independent and have a distribution F with mean $1/\lambda$. We make this somewhat unusual choice of notation for mean so that if $N(t)$ is the number of arrivals by time t, then our strong law, (2.1), implies that the long-run arrival rate is
$$\lim_{t \to \infty} \frac{N(t)}{t} = \frac{1}{Et_i} = \lambda$$
The second G stands for general service times. That is, we assume that the ith customer requires an amount of service s_i, where the s_i are independent and have a distribution G with mean $1/\mu$. Again, the notation for the mean is chosen so that the service rate is μ. The final 1 indicates there is one server.

Our first result states that the queue is stable if the arrival rate is smaller than the long-run service rate.

(3.1) Theorem. *Suppose $\lambda < \mu$. If the queue starts with some finite number $k \geq 1$ customers who need service, then it will empty out with probability one.*

Why is this true? While the queue is not empty the server works at rate μ, which is larger than the rate λ at which arrivals come, so the number of the people in the system tends to decrease and long queues will not develop.

Proof. We will proceed by contradiction using the idea above. Specifically we will show that if the server is always busy then since she serves customers faster than they arrive, she is eventually done with the customers before they arrive, which is a contradiction.

Turning to the details, let $T_n = t_1 + \cdots + t_n$ be the time of the nth arrival. The strong law of large numbers, (2.2), implies that

$$\frac{T_n}{n} \to \frac{1}{\lambda}$$

Let Z_0 be the sum of the service times of the customers in the system at time 0 and let s_i be the service time of the ith customer to arrive after time 0. If the server stays busy until the nth customer arrives then that customer will depart the queue at time $Z_0 + S_n$, where $S_n = s_1 + \cdots + s_n$. The strong law of large numbers implies

$$\frac{Z_0 + S_n}{n} \to \frac{1}{\mu}$$

Since $1/\mu < 1/\lambda$, this means that if we assume that the server is always working, then when n is large enough the nth customer departs before he arrives. This contradiction implies that the probability that the server stays busy for all time must be 0. □

By looking at the last argument more carefully we can conclude:

(3.2) Theorem. *Suppose $\lambda < \mu$. The limiting fraction of time the server is busy is λ/μ.*

Why is this true? Work arrives at rate λ. The server serves at rate μ but only can do work at a long-run rate of λ customers per unit time, since that is the rate at which work arrives. Thus the server must be busy at rate μ for a long-run fraction λ/μ of the time.

Section 5.3 Applications to Queueing Theory

More details. Suppose for simplicity that the queue starts out empty. The nth customer arrives at time $T_n = t_1 + \cdots + t_n$. If A_n is the amount of time the server has been busy up to time T_n and S_n is the sum of the first n service times, then

$$A_n = S_n - Z_n$$

where Z_n is the amount of work in the system at time T_n, i.e., the amount of time to empty the system if there were no more arrivals.

The first term is easy to deal with

$$\frac{S_n}{T_n} = \frac{S_n/n}{T_n/n} \to \frac{Es_i}{Et_i} = \frac{\lambda}{\mu}$$

As for the second term, we can argue intuitively that the condition $\lambda < \mu$ implies the queue reaches equilibrium, so EZ_n stays bounded, and we have $Z_n/n \to 0$. □

To prove (3.2) we would have to show that $Z_n/n \to 0$. To do this we can look at Y_n, which is the waiting time in the queue at the arrival of the nth customer, since $Z_n = Y_n + s_n$. We look at Y_n instead, since it satisfies the recursion

$(\star)$ $$Y_n = (Y_{n-1} + s_{n-1} - t_n)^+$$

where $z^+ = \max\{z, 0\}$. In words, the waiting time for the nth customer is waiting time for the $(n-1)$th plus the service time for the $(n-1)$th minus the interarrival time t_n since the server gets to work for an amount of time t_n. The positive part is needed in the formula since if the server gets to work for an amount of time larger than $Y_{n-1} + s_n$ then the queue empties out and $Y_n = 0$. Using the fact that $E(s_{n-1} - t_n) < 0$, it is possible to show that EY_n stays bounded and $Y_n/n \to 0$, but those details are too complicated to give here.

Cost equations. Let X_s be the number of customers in the system at time s. Let L be the long-run average number of customers in the system:

$$L = \lim_{t \to \infty} \frac{1}{t} \int_0^t X_s \, ds$$

Let W be the long-run average amount of time a customer spends in the system:

$$W = \lim_{n \to \infty} \frac{1}{n} \sum_{m=1}^n W_m$$

where W_m is the amount of time the mth arriving customer spends in the system. Finally, let λ_a be the long-run average rate at which arriving customers join the system, that is,
$$\lambda_a = \lim_{t \to \infty} N_a(t)/t$$
where $N_a(t)$ is the number of customers who arrive before time t and enter the system. Ignoring the problem of proving the existence of these limits, we can assert that these quantities are related by

(3.3) Little's formula.
$$L = \lambda_a W$$

Why is this true? Suppose each customer pays \$1 for each minute of time she is in the system. When ℓ customers are in the system, we are earning \$$\ell$ per minute, so in the long run we earn an average of \$$L$ per minute. On the other hand, if we imagine that customers pay for their entire waiting time when they arrive then we earn at rate $\lambda_a W$ per minute, i.e., the rate at which customers enter the system multiplied by the average amount they pay. □

For a simple example of the use of this formula consider:

Example 3.1. M/M/1 queue. Here arrivals are a rate λ Poisson process, there is one server, and customers require an amount of service that is exponentially distributed with mean $1/\mu$. If we assume $\lambda < \mu$, then it follows from Example 4.1 in Chapter 4 that the equilibrium probability of n people in the system is given by the shifted geometric distribution
$$\pi(n) = \left(1 - \frac{\lambda}{\mu}\right)\left(\frac{\lambda}{\mu}\right)^n$$
and the mean queue length in equilibrium is $L = \lambda/(\mu - \lambda)$. Since all customers enter the system $\lambda_a = \lambda$, and it follows that the average waiting time is
$$W = \frac{L}{\lambda} = \frac{1}{\mu - \lambda}$$
To check that this formula is sensible, note that (i) when λ is very small, this is close to $1/\mu$, the customer's service time; and (ii) when λ increases to μ, the average waiting time approaches infinity. □

For an example with $\lambda_a < \lambda$ we consider Example 3.3 of Chapter 4.

Example 3.2. Barbershop chain. A barber can cut hair at rate 3, where the units are people per hour, i.e., each haircut requires an exponentially distributed amount of time with mean 20 minutes. Customers arrive at times of a rate 2 Poisson process, but will leave if both chairs in the waiting room are full. We computed earlier that the equilibrium distribution for the number of people in the system was given by

$$\pi(0) = 27/65, \quad \pi(1) = 18/65, \quad \pi(2) = 12/65, \quad \pi(3) = 8/65$$

From this it follows that

$$L = 1 \cdot \frac{18}{65} + 2 \cdot \frac{12}{65} + 3 \cdot \frac{8}{65} = \frac{66}{65}$$

Customers will only enter the system if there are < 3 people, so

$$\lambda_a = 2(1 - \pi(3)) = 114/65$$

Combining the last two results and using Little's formula, (3.3), we see that the average waiting time in the system is

$$W = \frac{L}{\lambda_a} = \frac{66/65}{114/65} = \frac{66}{114} = 0.579 \text{ hours}$$

Example 3.3. Waiting time in the queue. Consider the $GI/G/1$ queue and suppose that we are only interested in the customer's average waiting time in the queue, W_Q. If we know the average waiting time W in the system, this can be computed by simply subtracting out the amount of time the customer spends in service

(3.4) $$W_Q = W - Es_i$$

For instance, in the previous example, subtracting off the 0.333 hours that his haircut takes we see that the customer's average time waiting in the queue $W_Q = 0.246$ hours or 14.76 minutes.

Let L_Q be the average queue length in equilibrium; i.e., we do not count the customer in service if there is one. If suppose that customers pay \$1 per minute in the queue and repeat the derivation of Little's formula, then

(3.5) $$L_Q = \lambda W_Q$$

The length of the queue is 1 less than the number in the system, except when the number in the system is 0, so if $\pi(0)$ is the probability of no customers, then

$$L_Q = L - 1 + \pi(0)$$

226 Chapter 5 Renewal Theory

Combining the last three equations with our first cost equation:

$$\pi(0) = L_Q - (L-1) = 1 + \lambda(W_Q - W) = 1 - \lambda E s_i$$

Recalling that $Es_i = 1/\mu$, we have another derivation of (3.2). □

The results above are valid for a general interarrival distribution. When the arrivals are a Poisson process we are able to get more detailed results:

b. M/G/1 queue

Here the M stands for Markovian input and indicates we are considering the special case of the $GI/G/1$ queue in which the inputs are a rate λ Poisson process. The rest of the set-up is as before: there is a one server and the ith customer requires an amount of service s_i, where the s_i are independent and have a distribution G with mean $1/\mu$.

When the input process is Poisson, the system has special properties that allow us to go further. We learned in (3.1) that if $\lambda < \mu$ then a $GI/G/1$ queue will repeatedly return to the empty state. Thus the server experiences alternating busy periods with duration B_n and idle periods with duration I_n. In the case of Markovian inputs, the lack of memory property implies that I_n has an exponential distribution with rate λ. Combining this observation with our result for alternating renewal processes we see that the limiting fraction of time the server is idle is

$$\frac{1/\lambda}{1/\lambda + EB_n} = 1 - \frac{\lambda}{\mu}$$

by (3.2). Rearranging, we have

$$(3.6) \qquad EB_n = \frac{1}{\lambda}\left(\frac{1}{1-(\lambda/\mu)} - 1\right) = \frac{1/\mu}{1-\lambda/\mu} = \frac{Es_i}{1-\lambda Es_i}$$

A second special property of Poisson arrivals is:

PASTA. These initials stand for "Poisson arrivals see time averages." To be precise, if $\pi(n)$ is the limiting fraction of time that there are n individuals in the queue and a_n is the limiting fraction of arriving customers that see a queue of size n, then

(3.7) Theorem. $a_n = \pi(n)$.

Observant readers will have noticed that we used this principle without mentioning it when we computed the waiting time distribution in the $M/M/1$ queue in Example 4.1 of Chapter 4. The result in (3.7) may sound obvious, but it is not true for a general arrival process. For a simple counterexample, suppose that the interarrival time is uniform on (3,5), while the service time is uniform on (0,2). If the queue starts empty, then since the mean time between arrivals is 4, and the mean service time is 1, our results for alternating renewal processes imply that $\pi(0) = 3/4$, $\pi(1) = 1/4$, but no arriving customer ever sees a queue so $a_0 = 1$.

Why is (3.7) true? If we condition on there being arrival at time t, then the times of the previous arrivals are a Poisson process with rate λ. Thus knowing that there is an arrival at time t does not affect the distribution of what happened before time t. □

For a simple application of (3.7) let us return to:

Example 3.5. Waiting times in the M/M/1 queue. Since each service time has mean $1/\mu$, it follows that

$$W = (1+L) \cdot \frac{1}{\mu} = \left(1 + \frac{\lambda}{\mu - \lambda}\right) \cdot \frac{1}{\mu} = \frac{1}{\mu - \lambda}$$

in agreement with the result of Example 3.1. □

Somewhat more sophisticated analysis gives information about:

Example 3.6. Workload in the M/G/1 queue. We define the workload in the system at time t, V_t, to be the sum of the remaining service times of all customers in the system. Suppose that each customer in the queue or in service pays at a rate of $\$y$ when his remaining *service* time is y; i.e., we do not count the remaining waiting time in the queue. If we let Y be the average total payment made by an arriving customer, then our cost equation reasoning implies that the average workload V satisfies

$$V = \lambda Y$$

(All customers enter the system, so $\lambda_a = \lambda$.) Since a customer with service time s_i pays s_i during the q_i units of time spent waiting in the queue and at rate $s_i - x$ after x units of time in service

$$Y = E(s_i q_i) + E\left(\int_0^{s_i} s_i - x \, dx\right)$$

Now a customer's waiting time in the queue can be determined by looking at the arrival process and at the service times of previous customers, so it is independent of her service time, i.e., $E(s_i q_i) = Es_i \cdot W_Q$ and we have

$$Y = (Es_i)W_Q + E(s_i^2/2)$$

PASTA implies that $V = W_Q$, so using $Y = V/\lambda$ and multiplying both sides by λ, we have

$$W_Q = \lambda(Es_i)W_Q + \lambda E(s_i^2/2)$$

Solving for W_Q now gives

(3.8) $$W_Q = \frac{\lambda E(s_i^2)}{2(1 - \lambda Es_i)}$$

the so-called **Pollaczek-Khintchine formula**. Using formulas (3.5), (3.4), and (3.3) we can now compute

$$L_Q = W_Q/\lambda \qquad W = W_Q + Es_i \qquad L = W/\lambda$$

5.4. Age and Residual Life

In this section we return to the study of renewal theory. Our goal is to show that, like Markov chains, these systems settle down to equilibrium as time tends to infinity. Renewal reward processes will be the key to our investigations.

Let $t_1, t_2, \ldots$ be i.i.d., let $T_n = t_1 + \cdots + t_n$ be the time of the nth renewal, and let $N(t) = \max\{n : T_n \leq t\}$ be the number of renewals by time t. Let

$$A(t) = t - T_{N(t)} \qquad \text{and} \qquad Z(t) = T_{N(t)+1} - t$$

$A(t)$ is the time since the last renewal before t, $Z(t)$ is the time until the next renewal after time t.

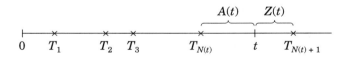

If we think in terms of light bulbs, $A(t)$ gives the age of the item in use at time t, while $Z(t)$ gives its residual lifetime.

Example 4.1. Long run average residual life. To study the asymptotic behavior of $Z(t)$, we begin by noting that when $T_{i-1} < s < T_i$, we have $Z(s) = T_i - s$ and changing variables $r = T_i - s$ gives

$$\int_{T_{i-1}}^{T_i} Z(s)\,ds = \int_0^{t_i} r\,dr = t_i^2/2$$

So ignoring the contribution from the last incomplete cycle, $[T_{N(t)}, t]$ we have

$$\int_0^t Z(s)\,ds \approx \sum_{i=1}^{N(t)} t_i^2/2$$

The right-hand side is a renewal reward process with $r_i = t_i^2/2$, so it follows from (2.1) that

(4.1) $$\frac{1}{t}\int_0^t Z(s)\,ds \to \frac{Et_i^2/2}{Et_i}$$

To see what this means we will consider two concrete examples:

A. Exponential. Suppose the t_i are exponential with rate λ. In this case $Et_i = 1/\lambda$, while integration by parts with $f(t) = t^2$ and $g'(t) = \lambda e^{-\lambda t}$ shows

$$Et_i^2 = \int_0^\infty t^2 \lambda e^{-\lambda t}\,dt = t^2(-e^{-\lambda t})\big|_0^\infty + \int_0^\infty 2t e^{-\lambda t}\,dt$$
$$= 0 + (2/\lambda)\int_0^\infty t\lambda e^{-\lambda t}\,dt = 2/\lambda^2$$

Using this in (4.1), it follows that

$$\frac{1}{t}\int_0^t Z(s)\,ds \to \frac{1/\lambda^2}{1/\lambda} = \frac{1}{\lambda}$$

This is not surprising since the lack of memory property of the exponential implies that for any s, $Z(s)$ has an exponential distribution with mean $1/\lambda$.

B. Uniform. Suppose the t_i are uniform on $(0, b)$, that is, the density function is $f(t) = 1/b$ for $0 < t < b$ and 0 otherwise. In this case the symmetry of the uniform distribution about $b/2$ implies $Et_i = b/2$, while a little calculus shows

$$Et_i^2 = \int_0^b t^2 \cdot \frac{1}{b}\,dt = \frac{t^3}{3b}\bigg|_0^b = \frac{b^2}{3}$$

Using this in (4.1) it follows that

$$\frac{1}{t}\int_0^t Z(s)\,ds \to \frac{b^2/6}{b/2} = \frac{b}{3}$$

Example 4.2. Long run average age. As in our analysis of $Z(t)$, we begin by noting that changing variables $s = T_{i-1} + r$ we have

$$\int_{T_{i-1}}^{T_i} A(s)\,ds = \int_0^{t_i} r\,dr = t_i^2/2$$

so ignoring the contribution from the last incomplete cycle $[T_{N(t)}, t]$ we have

$$\int_0^t A(s)\,ds \approx \sum_{i=1}^{N(t)} t_i^2/2$$

The right-hand side is the renewal reward process we encountered in Example 4.1, so it follows from (2.1) that

(4.2) $$\frac{1}{t}\int_0^t A(s)\,ds \to \frac{Et_i^2/2}{Et_i}$$

EXERCISE 4.1. Show that for a rate λ Poisson process, $EA(t) = (1 - e^{-\lambda t})/\lambda$.

Combining the results from the last two examples leads to a surprise.

Example 4.3. Inspection paradox. Let $L(t) = A(t) + Z(t)$ be the lifetime of the item in use at time t. Adding (4.1) and (4.2), we see that the average lifetime of the items in use up to time t:

(4.3) $$\frac{1}{t}\int_0^t L(s)\,ds \to \frac{Et_i^2}{Et_i}$$

To see that this is surprising, note that:

(i) If $\text{var}(t_i) = Et_i^2 - (Et_i)^2 > 0$, then $Et_i^2 > (Et_i)^2$, so the limit is $> Et_i$.

(ii) The average of the lifetimes of the first n items:

$$\frac{t_1 + \cdots + t_n}{n} \to Et_i$$

Thus, the average lifetime of items in use up to time t converges to a limit Et_i^2/Et_i, which is larger than Et_i, the limiting average lifetime of the first n items.

There is a simple explanation for this "paradox": taking the average age of the item in use up to time s is biased since items that last for time u are counted u times. For simplicity suppose that there are only a finite number of possible lifetimes $\ell_1 < \ell_2 \ldots < \ell_k$ with probabilities $p_1, \ldots, p_k > 0$ and $p_1 + \cdots + p_k = 1$.

By considering a renewal reward process in which $r_i = 1$ when $t_i = \ell_j$, we see that the number of items up to time t with lifetime ℓ_j is $\sim p_j N(t)$. The total length of these lifetimes is $\sim \ell_j p_j N(t)$ Thus the limiting fraction of time the lifetime is ℓ_j is by (2.1)

$$\sim \frac{\ell_j p_j N(t)}{t} \to \frac{\ell_j p_j}{Et_i}$$

The expected value of this limiting distribution is

$$\sum_j \ell_j \frac{\ell_j p_j}{Et_i} = \frac{Et_i^2}{Et_i}$$

Example 4.4. Limiting distribution for the residual life. Let $Z(t) = T_{N(t)+1} - t$. Let $I_c(t) = 1$ if $Z(t) \le c$ and 0 otherwise. To study the asymptotic behavior of $I_c(t)$, we begin with the observation that

$$\int_{T_{i-1}}^{T_i} I_c(s)\, ds = \min\{t_i, c\}$$

To check this we consider two cases.

Case 1. $t_i \ge c$. $I_c(s) = 1$ for $T_i - c \le s \le T_i$, 0 otherwise, so the integral is c.

Case 2. $t_i \le c$. $I_c(s) = 1$ for all $s \in [T_{i-1}, T_i]$, so the integral is t_i.

Ignoring the contribution from the last incomplete cycle $[T_{N(t)}, t]$, we have

$$\int_0^t I_c(s)\, ds \approx \sum_{i=1}^{N(t)} \min\{t_i, c\}$$

The right-hand side is a renewal reward process with $r_i = \min\{t_i, c\}$, so it follows from (2.1) that

(4.4) $$\frac{1}{t} \int_0^t I_c(s)\, ds \to \frac{E \min\{t_i, c\}}{Et_i}$$

Chapter 5 Renewal Theory

Example 4.5. Limiting distribution for the age. Let $A(t) = t - T_{N(t)+1}$ and let $J_c(t) = 1$ if $A(t) \le c$ and 0 otherwise. Imitating the last argument, it is easy to show that

$$(4.5) \qquad \frac{1}{t}\int_0^t J_c(s)\,ds \to \frac{E\min\{t_i, c\}}{Et_i}$$

Proof. To study the asymptotic behavior of $J_c(t)$ we begin with the observation that

$$\int_{T_{i-1}}^{T_i} J_c(s)\,ds = \min\{t_i, c\}$$

To check this we consider two cases.

Case 1. $t_i \ge c$. $J_c(s) = 1$ for $T_{i-1} \le s \le T_{i-1} + c$, 0 otherwise, so the integral is c.

Case 2. $t_i \le c$. $J_c(s) = 1$ for all $s \in [T_{i-1}, T_i]$, so the integral is t_i.

Ignoring the contribution from the last incomplete cycle $[T_{N(t)}, t]$, we have

$$\int_0^t I_c(s)\,ds \approx \sum_{i=1}^{N(t)} \min\{t_i, c\}$$

The right-hand side is a renewal reward process with $r_i = \min\{t_i, c\}$, so (4.5) follows from (2.1). □

To evaluate the limits in (4.4) and (4.5), we note that (1.9) implies

$$E\min\{t_i, c\} = \int_0^\infty P(\min\{t_i, c\} > t)\,dt = \int_0^c P(t_i > t)\,dt$$

From this it follows that the limiting fraction of time that the age of the item in use at time t is $\le c$ is

$$(4.6) \qquad G(c) = \frac{\int_0^c P(t_i > t)\,dt}{Et_i}$$

This is a distribution function of a nonnegative random variable since $G(0) = 0$, $t \to G(t)$ is nondecreasing, and $G(\infty) = 1$ by (1.9). Differentiating we find that the density function of the limiting age is given by

$$(4.7) \qquad g(c) = \frac{d}{dc}G(c) = \frac{P(t_i > c)}{Et_i}$$

Section 5.4 Age and Residual Life

Turning to our two concrete examples:

A. Exponential. In this case the limiting density given in (4.7) is

$$\frac{P(t_i > c)}{Et_i} = \frac{e^{-\lambda c}}{1/\lambda} = \lambda e^{-\lambda c}$$

For the residual life this is not surprising, since the distribution of $Z(s)$ is always exponential with rate λ.

B. Uniform. Plugging into (4.7) gives for $0 < c < b$:

(4.8)
$$\frac{P(t_i > c)}{Et_i} = \frac{(b-c)/b}{b/2} = \frac{2(b-c)}{b^2}$$

In words, the limiting density is a linear function that starts at $2/b$ at 0 and hits 0 at $c = b$.

In case A, the limit distribution $G = F$, while in case B, $G \neq F$. To show that case B is the rule to which case A is the only exception, we prove:

(4.9) Theorem. Suppose that $G = F$ in (4.6). Then $F(x) = 1 - e^{-\lambda x}$ for some $\lambda > 0$.

Proof. Let $H(c) = 1 - G(c)$, and $\lambda = 1/Et_i$. (4.6) implies $H'(c) = -\lambda H(c)$. Combining this with the observation $H(0) = 1$ and using the uniqueness of the solution of the differential equation, we conclude that $H(c) = e^{-\lambda c}$. □

Finally we get to the announced goal of the section:

Limiting behavior of renewal processes. Suppose that a renewal process has been running for a long time t. Let $T'_1, T'_2, \ldots$ be the amounts of time we have to wait for the first, second, etc., renewals after t. Let $t'_1 = T'_1$ be the waiting time to the first renewal after t and let $t'_k = T'_k - T'_{k-1}$ for $k \geq 2$ be the later inter renewal times. Results above imply that if t is large, then t'_1 will (approximately) have distribution G given in (4.6).

The distribution G is in general different from F since (i) the waiting time interval containing t is longer than usual and (ii) are we are only looking at the part of the waiting time that comes after t. However, the distribution of the later waiting times is unchanged.

(4.10) Lemma. t'_k, $k \geq 2$ are i.i.d. with distribution F are independent of t'_1.

Why is this true? $N = N(t) + 1$ is a stopping time, so applying the strong Markov property (3.1) in Chapter 1, to the random walk T_n, we see that $T_{N+j} -$

$T_N, j = 1, 2, \ldots$ is a random walk starting at 0 and is independent of $T_n, n \leq N$. Since $t'_k = T_{N+k-1} - T_{N+k-2}$ for $k \geq 2$, the desired result follows. □

The results above show that when t is large, the waiting times after t can be approximately described as follows. Let $u_1, u_2, \ldots$ be independent with u_1 having distribution G given in (4.7), and all the others having distribution F. Let $U_n = u_1 + \cdots + u_n$ and $M(t) = \max\{n : U_n \leq t\}$. A renewal process like $M(t)$ in which the initial waiting time for a renewal has a distribution than the remaining interrenewal times is called a **delayed renewal process**.

Here we will only be interested in the one in which G has the distribution given in (4.8). It is special because

(4.11) $$EM(t) = t/Et_i$$

(4.12)   If $Z(t) = U_{N(t)+1} - t$, then $Z(t)$ always has distribution G.

This delayed renewal process thus represents a stationary distribution for the original renewal process.

5.5. Exercises

Throughout these exercises we will use the notation of Chapter 5: $t_1, t_2, \ldots$ are independent and identically distributed interarrival times; $T_n = t_1 + \cdots + t_n$ is the time of the nth arrival; $N(t) = \max\{n : T_n \leq t\}$ is the number of arrivals by time t; and $m(t) = EN(t)$ is its expected value.

5.1. Suppose that the interarrival distribution in a renewal process is Poisson with mean μ. That is, suppose $P(t_i = k) = e^{-\mu}\mu^k/k!$. (a) Find the distribution of the time of the nth arrival time T_n. (b) Find $P(N(t) \geq n)$.

5.2. Continuing to use the notation of the previous problem, suppose that the interarrival times t_i distribution have a gamma$(2, \lambda)$ distribution, i.e., is the sum of two independent exponentials with rate λ. Find $P(N(t) = n)$.

5.3. Solve the previous problem for a gamma(m, λ) distribution.

5.4. Suppose that the interarrival times $t_1, t_2, \ldots$ are independent uniform $(0,1)$ random variables, let $T_n = t_1 + \cdots + t_n$ and $M = \min\{n : T_n > 1\}$. (a) What is EM? (b) Find $E(T_M - 1)$.

5.5. *Laplace transforms.* If we let $m(t) = EN(t)$, then differentiating (1.4) gives $m'(t) = \sum_{n=1}^{\infty} P(T_n = t)$. Multiply by $e^{-\theta t}$ and integrate to conclude

$$\int_0^\infty e^{-\theta t} m'(t)\, dt = \frac{\varphi(\theta)}{1 - \varphi(\theta)}$$

5.6. Suppose the interarrival times in a renewal process follow a gamma(2,1) density. That is, $P(t_i = t) = te^{-t}$ for $t > 0$. (a) Use the previous exercise to conclude

$$\int_0^\infty e^{-\theta t} m'(t)\, dt = \frac{1}{\theta(\theta+2)} = \frac{1}{2\theta} - \frac{1}{2(\theta+2)}$$

(b) Verify that

$$\int_0^\infty e^{-\theta s}\left(\frac{1}{2} - \frac{1}{2}e^{-2s}\right) ds = \frac{1}{2\theta} - \frac{1}{2(\theta+2)}$$

Since this holds for all θ it follows from results about Laplace transforms that $m'(s) = (1 - e^{-2s})/2$. Now integrate to find $m(t)$.

5.7. Suppose that the interarrival times in a renewal process have density function

$$\frac{1}{2}\cdot e^{-t} + \frac{1}{2}\cdot 3e^{-3t}$$

(a) Compute the Laplace transform of this density function. (b) Use Exercise 5.5 to conclude

$$\int_0^\infty e^{-\theta t} m'(t)\, dt = \frac{3 + 2\theta}{\theta(\theta+2)}$$

(c) Use the methods of the previous exercise to conclude that

$$m(t) = \frac{3t}{2} - \frac{1}{4}\left(e^{-2t} - 1\right)$$

(d) Why does having 1/2 of the arrivals at rate 1 and 1/2 at rate 3 lead to an asymptotic rate of 3/2 instead of 2?

5.8. Suppose the interarrival times are a general mixture of two exponential distributions:

$$P(t_i = t) = p\mu_1 e^{-\mu_1 t} + (1-p)\mu_2 e^{-\mu_2 t}$$

Use the methods of the previous problem to show $m(t) = At + C(e^{-rt} - 1)$, then find A and r.

5.9. The weather in a certain locale consists of alternating wet and dry spells. Suppose that the number of days in each rainy spell is a Poisson distribution with mean 2, and that a dry spell follows a geometric distribution with mean 7. Assume that the successive durations of rainy and dry spells are independent. What is the long-run fraction of time that it rains?

5.10. Monica works on a temporary basis. The mean length of each job she gets is 11 months. If the amount of time she spends between jobs is exponential with mean 3 months, then in the long run what fraction of the time does she spend working?

5.11. Thousands of people are going to a Greatful dead concert in Pauley Pavillion at UCLA. They park their 10 foot cars on several of the long streets near the arena. There are no lines to tell the drivers where to park, so they park at random locations, and end up leaving spacings between the cars that are independent and uniform on $(0,10)$. In the long run, what fraction of the street is covered with cars?

5.12. The times between the arrivals of customers at a taxi stand are independent and have a distribution F with mean μ_F. Assume an unlimited supply of cabs, such as might occur at an airport. Suppose that each customer pays a random fare with distribution G and mean μ_G. Let $W(t)$ be the total fares paid up to time t. Find $\lim_{t \to \infty} EW(t)/t$.

5.13. A policeman cruises (on average) approximately 10 minutes before stopping a car for speeding. 90% of the cars stopped are given speeding tickets with an \$80 fine. It takes the policeman an average of 5 minutes to write such a ticket. The other 10% of the stops are for more serious offenses, leading to an average fine of \$300. These more serious charges take an average of 30 minutes to process. In the long run, at what rate does he assign fines.

5.14. A group of n children continuously, and independently, climb up and then sled down a slope. Assume that each child's actions follow an alternating renewal process: climbing up for an amount of time with distribution F and mean μ_F, and then sliding down for an amount of time with distribution G and mean μ_G. Let $U(t)$ be the number of children climbing the hill at time t. Find $\lim_{t \to \infty} P(U(t) = k)$.

5.15. *Counter processes.* As in Example 1.5, we suppose that arrivals at a counter come at times of a Poisson process with rate λ. An arriving particle that finds the counter free gets registered and then locks the counter for an amount of time τ. Particles that arrive while the counter is locked have no effect. (a) Find the limiting probability the counter is locked at time t. (b) Compute the limiting fraction of particles that get registered.

5.16. A cocaine dealer is standing on a street corner. Customers arrive at times of a Poisson process with rate λ. The customer and the dealer then disappear from the street for an amount of time with distribution G while the transaction is completed. Customers that arrive during this time go away never to return. (a) At what rate does the dealer make sales? (b) What fraction of customers are lost?

5.17. A worker has a number of machines to repair. Each time a repair is completed a new one is begun. Each repair independently takes an exponential amount of time with rate μ to complete. However, independent of this, mistakes occur according to a Poisson process with rate λ. Whenever a mistake occurs, the item is ruined and work is started on a new item. In the long run how often are jobs completed?

5.18. Three children take turns shooting a ball at a basket. The first shoots until she misses, then the second shoots until she misses, the third shoots until she misses, then the process starts again with the first child. Suppose that child i makes a basket with probability p_i and that successive trials are independent. Determine the proportion of time in the long run that each child shoots.

5.19. Solve the previous problem when $p_1 = 2/3$, $p_2 = 3/4$, $p_3 = 4/5$.

5.20. Bruno works for an amount of time that is uniformly distributed on $[6, 10]$ hours then he relaxes for an exponentially distributed number of hours; with mean 4. (a) Bruno never sleeps. Ignoring mundane details like night and day, what is the long-run fraction of time he spends working? (b) Suppose that when Bruno first begins relaxing he drinks a beer and then drinks one beer each hour after that. That, is if his relaxation period is from 12:13 to 3:27 he drinks 4 beers, one each at 12:13, 1:13, 2:13, and 3:13. Find the probability he drinks n beers in one relaxation period and the long-run average number of beers he drinks each 24 hours.

5.21. A young doctor is working at night in an emergency room. Emergencies come in at times of a Poisson process with rate 0.5 per hour. The doctor can only get to sleep when it has been 36 minutes (.6 hours) since the last emergency. For example, if there is an emergency at 1:00 and a second one at 1:17 then she will not be able to get to sleep until at least 1:53, and it will be even later if there is another emergency before that time. (a) Compute the long-run fraction of time she spends sleeping, by formulating a renewal reward process in which the reward in the ith interval is the amount of time she gets to sleep in that interval. (b) The doctor alternates between sleeping for an amount of time s_i and being awake for an amount of time u_i. Use the result from (a) to compute Eu_i.

5.22. A scientist has a machine for measuring ozone in the atmosphere that is located in the mountains just north of Los Angeles. At times of a Poisson process with rate 1, storms or animals disturb the equipment so that it can no longer collect data. The scientist comes every L units of time to check the equipment. If the equipment has been disturbed then she can usually fix it quickly so we will assume the the repairs take 0 time. (a) What is the limiting fraction of time the machine is working? (b) Suppose that the data that is being collected is worth a dollars per unit time, while each inspection costs $c < a$. Find the best value of the inspection time L.

5.23. A man buys a new car planning to keep it for T years. The car has an exponential lifetime with mean 5 years. If the car fails before time T, it goes to the junkyard and he must immediately buy a new one. If the car survives to time T, then he can sell it for half the original price and buy a new one. To make the arithmetic simple, assume that cars cost \$4. (a) What is the average cost per year of this strategy? (b) Show that the optimal strategy is to never sell the car. (c) Give an intuitive explanation for the result in (b).

5.24. Suppose a new car has a lifetime that is uniform on $(2,8)$. A man buys a new car planning to keep it for T years. Again suppose that new cars cost \$4. If the car fails before time T, a cost of \$1 is incurred. If the car survives to time T then he can sell it for $R(T) = (8-T)/2$. (a) Suppose that $2 \le T \le 8$. What is the average cost per year of this strategy? (b) Find the optimal value of T.

5.25. Customers arrive at a full-service one-pump gas station at rate 20 cars per hour. However, customers will go to another station if there are at least two cars in the station, i.e., one being served and one waiting. Suppose that the service time for customers is exponential with mean 6 minutes. (a) Find the stationary distribution for the number of cars at the gas station. (b) Use Little's formula to compute the average waiting time for someone who enters the system. (c) What is the average waiting time in the queue? (d) Check the answer to (c) by noting that an arriving customer has to wait if and only if there is 1 car in the system.

5.26. Two people who prepare tax forms are working in a store at a local mall. Each has a chair next to his desk where customers can sit and be served. In addition there is one chair where customers can sit and wait. Customers arrive at rate 2 but will go away if there is already someone sitting in the chair waiting. Suppose that each server requires an exponential amount of time with rate 3 to complete the tax form. In Exercise 8.10 of Chapter 4 we formulated a Markov chain model for this system with state space $\{0, 1, 2, 12, 3\}$ where the first four states indicate the servers that are busy while the last indicates that there is a total of three customers in the system: one at each server and one waiting. (a) Find the stationary distribution. (b) Use Little's formula to compute the average waiting time of a customer who enters the system. (c) What is the average waiting time in the queue? (d) Check the answer to (c) by noting that an arriving customer has to wait if and only if there is 1 car in the system.

5.27. There are two tennis courts. Pairs of players arrive at rate 3 per hour and play for an exponentially distributed amount of time with mean 1 hour. If there are already two pairs of players waiting they will leave. (a) Find the stationary distribution for the number of courts occupied. (b) Use Little's formula to compute the average time spent waiting (i.e., not playing) for players who enter the system.

5.28. In Example 3.4 of Chapter 4 we considered a factory that has three machines in use and one repairman. We supposed that each machine worked for an exponential amount of time with mean 60 days between breakdowns, but each breakdown requires an exponential repair time with mean 4 days. We found that the stationary distribution for the number of broken machines was

$$\pi(0) = \frac{1225}{1480} \quad \pi(1) = \frac{225}{1480} \quad \pi(2) = \frac{30}{1480} \quad \pi(3) = \frac{2}{1480}$$

(a) Suppose the three machines are colored red, yellow, and blue. Use the stationary distribution to compute the long run fraction of time the red machine is in the repair shop. (b) A given machine alternates between being in the repair shop for time s_i and then working for an amount of time u_i. Use the answer to (a) to compute Es_i. (c) Is the process considered in (b) an alternating renewal process?

5.29. The city of Ithaca, New York, allows for two-hour parking in all downtown spaces. Methodical parking officials patrol the downtown area, passing the same point every two hours. When an official encounters a car, he marks it with chalk. If the car is still there two hours later, a ticket is written. Suppose that you park your car for a random amount of time that is uniformly distributed on $(0, 4)$ hours. What is the probability you will get a ticket.

5.30. Each time the frozen yogurt machine at the mall breaks down, it is replaced by a new one of the same type. In the long run what percentage of time is the machine in use less than one year old if the the lifetime distribution of a machine is: (a) uniformly distributed on (0,2)? (b) exponentially distributed with mean 1?

5.31. While visiting Haifa, Sid Resnick discovered that people who wish to travel quickly from the port area up the mountain to The Carmel frequently take a taxi known as a sherut. The system operates as follows: Sherut-eem are lined up in a row at a taxi stand. The capacity of each car is 5 people. Potential customers arrive according to a Poisson process with rate λ. As soon as 5 people are in the car, it departs in a cloud of diesel emissions. The next car moves up, accepts passengers until it is full, then departs for The Carmel, and so on. A local resident (who has no need of a ride) wanders onto the scene. What is the distribution of the time he has to wait to see a cab depart?

5.32. In front of terminal C at the Chicago airport is an area where hotel shuttle vans park. Customers arrive at times of a Poisson process with rate 10 per hour looking for transportation to the Hilton hotel nearby. When 7 people are in the van it leaves for the 36-minute round trip to the hotel. Customers who arrive while the van is gone go to some other hotel instead. (a) What fraction of the customers actually go to the Hilton? (b) What is the average amount of time that a person who actually goes to the Hilton ends up waiting in the van?

5.33. Let H be an increasing function with $H(0) = 0$ and $h(x) = H'(x)$. Generalize the proof of (1.2) to conclude that if $X \geq 0$, then

$$EH(X) = \int_0^\infty h(t) P(X > t) \, dt$$

In particular, $EX^2 = \int_0^\infty 2t P(X > t) \, dt$.

5.34. Let $t_1, t_2, \ldots$ be independent and identically distributed with $Et_i = \mu$ and $\text{var}(t_i) = \sigma^2$. Use the result of the previous exercise to compute the mean of the limiting distribution for the residual lifetime given in (4.7).

5.35. Let $A(t)$ and $Z(t)$ be, respectively, the age and the residual life at time t in a renewal process in which the interarrival times have distribution F. Compute $P(Z(t) > x | A(t) = s)$.

5.36. Let $A(t)$ and $Z(t)$ be, respectively, the age and the residual life at time t in a renewal process in which the interarrival times have distribution F. Use the methods of Section 5.4 to compute the limiting behavior of the joint distribution $P(A(t) > x, Z(t) > y)$.

5.37. A game is played on a very long and simple-looking board with squares 0 = start, 1, 2, ... On each turn you roll one die (with sides numbered 1, 2, 3, 4, 5, 6) and then move forward that number of spaces. To make it easier to state the next two questions, we will suppose that when your marker comes to rest on a square it leaves a black dot. (a) Let N_n be the number of squares with numbers $\leq n$ that you have visited (i.e., are marked with black dots). Find the limiting behavior of the fraction of squares you visit N_n/n. (b) Let X_n be a Markov chain with state space $\{0, 1, 2, 3, 4, 5\}$ so that

If $X_n > 0$, then $X_{n+1} = X_n - 1$.

If $X_n = 0$, then $X_{n+1} = 0, 1, 2, 3, 4, 5$ with probability $1/6$ each.

X_n is the distance from square n to the next square $m \geq n$ marked with a black dot. Find $\pi_i = \lim_{n \to \infty} P(X_n = i)$. Note that $X_n = 0$ if we visited that square, so π_0 should agree with your answer to (a).

5.38. *Discrete renewal theory.* Generalizing from the previous exercise, let $t_1, t_2, \ldots$ be independent integer valued random variables with $P(t_i = m) = p_m$ and mean $\mu = Et_i < \infty$. Let $T_k = t_1 + \cdots + t_k$ be the time of the kth renewal and let

$$X_n = \min\{T_k - n : T_k \geq n\}$$

be the time until the next renewal $\geq n$. It is easy to see that X_n is a Markov chain. (a) Compute its transition probability. (b) Let

$$\pi(i) = \frac{1}{\mu} \sum_{m=i+1}^{\infty} p_m$$

Show that π is a stationary distribution.

6 Brownian Motion

6.1. Basic Definitions

Brownian motion is the most important example of a process with continuous time and continuous paths. Mathematically, it lies in the intersection of three classes of stochastic processes: it is a Gaussian process, a Markov process with continuous paths, and has independent increments. Because of this, it is possible to use techniques from all three fields, so there is a rich and detailed theory of its properties. From a practical view, Brownian motion is often used as a component of models of physical, biological, and economic phenomena.

(1.1) Definition. $B(t)$ is a (one-dimensional) Brownian motion with variance σ^2 if it satisfies the following conditions:

(a) $B(0) = 0$, a convenient normalization.

(b) *Independent increments.* Whenever $0 = t_0 < t_1 < \ldots < t_n$

$$B(t_1) - B(t_0), \ldots, B(t_n) - B(t_{n-1}) \quad \text{are independent.}$$

(c) *Stationary increments.* The distribution of $B_t - B_s$ only depends on $t - s$.

(d) $B(t)$ is normal$(0, \sigma^2 t)$.

(e) $t \to B_t$ is continuous.

To explain the definition we begin by discussing the situation that gives the process its name. Historically, Brownian motion originates from the fact observed by Robert Brown (and others) in the 1800s that pollen grains under a microscope perform a "continuous swarming motion." To explain the source of this motion, we imagine a spherical pollen grain that collides with water molecules at times of a Poisson process with a large rate λ. The displacement of the pollen grain between time 0 and time 1 is the sum of a large number of small independent random effects. To see that this implies the change in the x coordinate, $X(1) - X(0)$, will be approximately normal$(0, \sigma^2)$, we recall:

(1.2) Central Limit Theorem. Let $X_1, X_2, \ldots$ be i.i.d. with $EX_i = 0$ and var$(X_i) = \sigma^2$, and let $S_n = X_1 + \cdots + X_n$. Then for all x we have

$$P\left(\frac{S_n}{\sigma\sqrt{n}} \leq x\right) \to P(\chi \leq x)$$

where χ has a standard normal distribution. That is,

$$P(\chi \leq x) = \int_{-\infty}^{x} \frac{1}{\sqrt{2\pi}} e^{-y^2/2} \, dy$$

Formally, (1.2) says that $S_n/\sigma\sqrt{n}$ **converges in distribution** to χ, a conclusion we will write as $S_n/\sigma\sqrt{n} \Rightarrow \chi$.

Since there are a huge number of water molecules in the system ($> 10^{20}$), it is reasonable to think that after a collision the molecule disappears into the crowd never to be seen again, and with a small leap of faith we can assert that $X(t)$ will have independent increments. The movement of the pollen grain is clearly continuous in space, so if we impose the condition $X(0) = 0$ by measuring displacements from the initial position, then the first coordinate of the pollen grain will be a Brownian motion with variance σ^2.

Does Brownian motion exist? Is there a process with the properties given in (1.1)? The answer is yes, of course, or this chapter wouldn't exist. The technical problem is to define the uncountably many random variables B_t for each nonnegative real number t on the same space so that $t \to B_t$ is continuous with probability one. One solution to this problem is provided in Exercise 6.11. However, the details of the proof are not important for understanding how B_t behaves, so we suggest that the reader take the existence of Brownian motion on faith.

Multidimensional Brownian motion. If we are looking at the pollen grain through a microscope then we can observe two coordinates of its motion $(X^1(t), X^2(t))$. Again we can argue that the displacement of the pollen grain between time 0 and time t will be the sum of a large number of small independent random effects, so if we suppose that $(X^1(0), X^2(0)) = (0, 0)$, then the position at time t, $(X^1(t), X^2(t))$, will have a two-dimensional normal distribution.

Multidimensional normal distributions $(Z^1, \ldots, Z^n)$ are characterized by giving the vector of means $\mu_i = EZ^i$ and the covariance matrix:

$$\Gamma_{ij} = E\{(Z^i - \mu_i)(Z^j - \mu_j)\}$$

244 Chapter 6 Brownian Motion

In the case under consideration $\mu_1 = \mu_2 = 0$. The spherical symmetry of the pollen grain and of the physical system implies that

$$\Gamma_{ij} = \sigma^2 I_{ij}$$

where I is the identity matrix: $I_{ii} = 1$ and $I_{ij} = 0$ for $i \neq j$. Recalling that:

(1.3) Lemma. *Two components Z_i and Z_j of a multidimensional normal are independent if and only if their covariance $E(Z_i Z_j) - EZ_i EZ_j = 0$*

we see the two components of the movement of the pollen grain are independent. Generalizing (and specializing), we state:

(1.4) Definition. *$(B^1(t), \ldots B^n(t))$, $t \geq 0$ is a standard n-dimensional Brownian motion if each of its components are independent one-dimensional Brownian motions with $\sigma^2 = 1$.*

We have removed the parameter σ^2 since it is trivial to reintroduce it. If B_t is a standard Brownian motion, then σB_t gives a Brownian motion with variance σ^2. A slight variation on the last calculation gives the following useful result:

(1.5) Scaling relation. *The processes $\{B_{ct}, t \geq 0\}$ and $\{c^{1/2} B_t, t \geq 0\}$ have the same distribution.*

Proof. Speeding up time by a factor of c does not change properties (a), (b), (c), or (e), but in (d) it makes the variance $c(t-s)$. By the previous observation this can be achieved by multiplying a standard Brownian motion by $c^{1/2}$. □

A second basic property of Brownian motion is:

(1.6) Covariance formula. $E(B_s B_t) = s \wedge t$.

Proof. If $s = t$, this says $EB_s^2 = s$, which is correct since B_s is normal$(0, s)$, so suppose now that $s < t$. Writing $B_t = B_s + (B_t - B_s)$ and using the fact that B_s and $B_t - B_s$ are independent with mean 0, we have

$$E(B_s B_t) = E(B_s^2) + E(B_s(B_t - B_s)) = s + 0 = s = s \wedge t \quad \square$$

It follows easily from the definition given in (1.1) that Brownian motion is a **Gaussian process**. That is, for any $t_1 < t_2 \ldots < t_n$ the vector

$$(B(t_1), B(t_2), \ldots B(t_n))$$

has a multivariate normal distribution. Since multivariate normal distributions are characterized by giving their means and covariances, it follows that Brownian motion is the only Gaussian process with continuous paths that has $EB_s = 0$ and $E(B_s B_t) = s \wedge t$.

We will not use the Gaussian viewpoint here, except in a couple of the exercises. However, to complete the story we should mention two important Gaussian relatives of Brownian motion:

Example 1.1. Brownian bridge. Let $B_t^0 = B_t - tB_1$. By definition $B_1^0 = 0$. The basic fact about this process is the following:

(1.7) Theorem. $\{B_t^0, 0 \le t \le 1\}$ and $(\{B_t, 0 \le t \le 1\} | B_1 = 0)$ have the same distribution. Each is a Gaussian process with mean 0 and covariance $s(1-t)$ if $s \le t$.

Why is this true? Since each of the processes is Gaussian and has mean 0, it suffices to show that their covariances are equal to $s(1-t)$. If $s < t$,

$$E(B_s^0 B_t^0) = E((B_s - sB_1)(B_t - tB_1)) = EB_s B_t - sEB_1 B_t - tEB_s B_1 + st B_1^2$$
$$= s - st - st + st = s(1-t)$$

With considerably more effort (see Exercise 6.2) one can compute the covariance function for $(B_t | B_1 = 0)$ and show that it is also $s(1-t)$, so the two processes have the same distribution. □

Example 1.2. Ornstein–Uhlenbeck process. Let

$$Z(t) = e^{-t} B(e^{2t}) \quad \text{for } -\infty < t < \infty$$

The definition may look strange, but note the very special property that for each t, $Z(t)$ has a normal(0,1) distribution. To compute the joint distribution of $Z(t)$ and $Z(t+s)$ we note that

$$Z(s+t) = e^{-(s+t)} B(e^{2(s+t)})$$
$$= e^{-(s+t)} B(e^{2t}) + e^{-(s+t)}(B(e^{2(s+t)}) - B(e^{2t}))$$

The first term is equal to $e^{-s} Z(t)$. The second is independent of $Z(t)$ and has a normal distribution with variance $e^{-2(s+t)}(e^{2(s+t)} - e^{2t}) = 1 - e^{-2s}$. Thus introducing an independent standard normal χ we can write

(⋆) $$Z(s+t) = e^{-s} Z(t) + \chi \sqrt{1 - e^{-2s}}$$

From this we can see that the effect of $Z(t)$ decays exponentially in time and that as $s \to \infty$, $Z(s+t)$ is almost an independent standard normal. Using ($\star$) we can also compute the covariance for of the Ornstein–Uhlenbeck process: $E(Z(t)Z(t+s)) = e^{-s} EZ(t)^2 + 0 = e^{-s}$.

6.2. Markov Property, Reflection Principle

Intuitively, the fact that Brownian motion has stationary independent increments (i.e., (c) and (b) of the definition hold) should immediately imply:

(2.1) Markov property. If $s > 0$, then $B_{s+t} - B_s$, $t \geq 0$ is a Brownian motion independent of B_r, $r \leq s$.

In words, the future increments $B_{s+t} - B_s$, $t \geq 0$ are independent of "the behavior of the process before time s." The phrase in quotation marks is intuitive but a little slippery to articulate mathematically. To be precise we should say instead of (2.1) that

(2.1') For any times $0 \leq r_1 \cdots < r_n = s$, $B_{s+t} - B_s$, $t \geq 0$ is a Brownian motion independent of the vector $(B(r_1), \ldots, B(r_n))$.

However, we prefer the more informal phrase "independent of $\{B_r, r \leq s\}$" so we will say the short phrase and ask that the reader understand that what we mean precisely is the longer one given in (2.1').

As in discrete time, the Markov property is also true at **stopping times** T which are defined by the requirement that the decision to stop before time s, $T \leq s$, can be determined from the values of B_r at times $r \leq s$.

(2.2) Strong Markov property. If T is a stopping time, then $B_{T+t} - B_T$, $t \geq 0$ is a Brownian motion independent of B_r, $r \leq T$.

The proof of (2.2), or even a precise definition of the notion of stopping time, requires considerable sophistication. Historically, it was not given until many years after Brownian motion was first used as a model. Here, we will follow early researchers and use (2.2) without bothering about its proof.

To illustrate the use of (2.2), let $T_a = \min\{t \geq 0 : B_t = a\}$ be the first time that Brownian motion hits a. Our first fact is:

(2.3) T_a, $a \geq 0$ **has stationary independent increments.** That is, (i) if $a < b$, then the distribution of $T_b - T_a$ is the same as T_{b-a}. (ii) if $a_0 = 0 < a_1 < \cdots < a_n$, then $T_{a_1} - T_{a_0}, \ldots, T_{a_n} - T_{a_{n-1}}$ are independent.

Section 6.2 Markov Property, Reflection Principle

Proof. Using (2.2) with $T = T_a$ we see that $B_{T+t} - B_T$ is a Brownian motion independent of B_r, $r \leq T$. Since $B_T = a$, we see that $T_b - T_a$ is just the hitting time of $b - a$ for $B_{T+t} - B_T$ and hence the distribution of $T_b - T_a$ is the same as T_{b-a}.

To prove (ii) now, we use (2.2) with $T = T_{a_{n-1}}$ to see that $B_{T+t} - B_T$ is a Brownian motion independent of B_r, $r \leq T$. Since $T_{a_n} - T_{a_{n-1}}$ is determined by $B_{T+t} - B_T$ for $t \geq 0$, while $(T_{a_1} - T_{a_0}, \ldots, T_{a_{n-1}} - T_{a_{n-2}})$ is determined by B_r, $r \leq T$, we see that the last increment $T_{a_n} - T_{a_{n-1}}$ is independent of the previous $n - 1$ increments. This result and an induction argument shows that the increments are independent. □

Distribution of T_a. Suppose $a > 0$. Since $P(B_t = a) = 0$ we have

(α) $$P(T_a \leq t) = P(T_a \leq t, B_t > a) + P(T_a \leq t, B_t < a)$$

Now any path that starts at 0 and ends above $a > 0$ must cross a, so

(β) $$P(T_a \leq t, B_t > a) = P(B_t > a)$$

To handle the second term in (α), we note that (i) the strong Markov property implies that $B(T_a+t) - B(T_a) = B(T_a+t) - a$ is a Brownian motion independent of B_r, $r \leq T_a$, and (ii) the normal distribution is symmetric about 0, so

(γ) $$P(T_a \leq t, B_t < a) = P(T_a \leq t, B_t > a)$$

The last argument is sometimes called the **reflection principle** since it was originally obtained by arguing that a path that started at 0 and ended above a had the same probability as the one obtained by reflecting the path B_t, $t \geq T_a$ is an mirror located at level a.

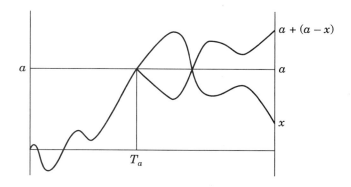

248 Chapter 6 Brownian Motion

Combining (α), (β), and (γ), we have

(2.4) $$P(T_a \le t) = 2P(B_t > a) = 2\int_a^\infty (2\pi t)^{-1/2} \exp(-x^2/2t)\, dx$$

To find the probability density of T_a, we change variables $x = t^{1/2}a/s^{1/2}$, with $dx = -t^{1/2}a/2s^{3/2}ds$ to get

$$P_0(T_a \le t) = 2\int_t^0 (2\pi t)^{-1/2} \exp(-a^2/2s)\left(-t^{1/2}a/2s^{3/2}\right) ds$$

$$= \int_0^t (2\pi s^3)^{-1/2} a \exp(-a^2/2s)\, ds$$

Thus we have shown:

(2.5) T_a has density function

$$(2\pi s^3)^{-1/2} a \exp(-a^2/2s)$$

Using (2.5) and integrating, we can show:

(2.6) The probability that B_t has no zero in the interval (s, t) is

$$\frac{2}{\pi} \arcsin\sqrt{\frac{s}{t}}$$

Proof. If $B_s = a > 0$, then symmetry implies that the probability of at least one zero in (s, t) is $P(T_a \le t - s)$. Breaking things down according to the value of B_s, which has a normal$(0, s)$ distribution, and using the Markov property, it follows that the probability of interest is

$$2\int_0^\infty \frac{1}{\sqrt{2\pi s}} e^{-a^2/2s} \int_0^{t-s} (2\pi r^3)^{-1/2} a \exp(-a^2/2r)\, dr\, da$$

Interchanging the order of integration, we have

(*) $$\frac{1}{\pi\sqrt{s}} \int_0^{t-s} r^{-3/2} \int_0^\infty a \exp\left(-a^2/2 \left[\frac{1}{s} + \frac{1}{r}\right]\right) da\, dr$$

The inside integral can be evaluated as

$$= \frac{sr}{r+s} \exp\left(-\frac{a^2 rs}{2(r+s)}\right)\bigg|_0^\infty = \frac{sr}{r+s}$$

so the expression in (∗) is equal to

$$\frac{\sqrt{s}}{\pi} \int_0^{t-s} \frac{1}{(s+r)\sqrt{r}}\, dr$$

Changing variables $r = sv^2$, $dr = 2sv\, dv$ produces

$$\frac{2}{\pi} \int_0^{\sqrt{(t-s)/s}} \frac{dv}{1+v^2} = \frac{2}{\pi} \arctan\sqrt{\frac{t-s}{s}}$$

By considering a right triangle with sides $\sqrt{s}$, $\sqrt{t-s}$ and $\sqrt{t}$ one can see that when the tangent is $\sqrt{(t-s)/s}$, the cosine is $\sqrt{s/t}$. At this point we have shown that the probability of at least one zero in (s,t) is $(2/\pi)\arccos(\sqrt{s/t})$. To get the result given, recall $\arcsin(x) = (\pi/2) - \arccos(x)$ when $0 \leq x \leq 1$. □

Let $L_t = \max\{s < t : B_s = 0\}$. From (2.6) we see that

(2.7) $$P(L_t < s) = \frac{2}{\pi} \arcsin\sqrt{\frac{s}{t}}$$

Recalling that the derivative of $\arcsin(x)$ is $1/\sqrt{1-x^2}$, we see that L_t has density function

(2.8) $$P(L_t = s) = \frac{2}{\pi} \frac{1}{\sqrt{1-s/t}} \cdot \frac{1}{2\sqrt{st}} = \frac{1}{\pi\sqrt{s(t-s)}}$$

In view of the definition of L_t as the LAST ZERO before time t, it is somewhat surprising that the density function of L_t

(i) is symmetric about $t/2$, and

(ii) tends to ∞ as $s \to 0$.

In a different direction, we also find it remarkable that $P(L_t > 0) = 1$. To see why this is surprising note that if we change our previous definition of the hitting time of 0 to $T_0^+ = \min\{s > 0 : B_s = 0\}$ then this implies that for all $t > 0$ we have $P(T_0^+ \leq t) = P(L_t > 0) = 1$, and hence

(2.9) $$P(T_0^+ = 0) = 1$$

Since T_0^+ is defined to be the minimum over all strictly positive times, (2.9) implies that there is a sequence of times $s_n \to 0$ at which $B(s_n) = 0$. At first this may sound innocent, but using this observation with the strong Markov property, we see that every time Brownian motion returns to 0 it will immediately have infinitely many 0's!

6.3. Martingales, Hitting Times

As in our earlier study of random walks in Section 2.4, martingales are useful for proving results about Brownian motion. The first and simplest is the:

a. Linear martingale.

(3.1) Theorem. B_t is a martingale. That is, if $s < t$, then

$$E(B_t|B_r, r \leq s) = B_s$$

What does this formula mean? In words, it says that the conditional expectation of B_t given the entire history of the process before time s is equal to the value at time s. As was the case with the Markov property, it is difficult to define formally the conditional expectation given the entire past, so as in the discussion of the Markov property in Section 6.2, what this really means is that

(3.1′) For any times $0 \leq r_1 \cdots < r_n < s$,

$$E(B_t|B_s, B_{r_n}, \ldots, B_{r_1}) = B_s$$

Again, we prefer the more informal formulation of the conditional expectation as "given B_r, $r \leq s$" rather than the technically correct "given $B_s, B_{r_n}, \ldots, B_{r_1}$ for any times $0 \leq r_1 \cdots < r_n < s$." Again, we will say the short phrase and ask the reader to remember that what we mean precisely is the long one.

Proof of (3.1). Given B_r, $r \leq s$, the value of B_s is known while $B_t - B_s$ is independent and has mean 0.

$$E(B_t|B_r, r \leq s) = E(B_s + B_t - B_s|B_r, r \leq s)$$
$$= B_s + E(B_t - B_s|B_r, r \leq s) = B_s \qquad \square$$

As in Chapter 2, all of our computations will be based on the:

(3.2) Stopping theorem for bounded martingales. Let M_t, $t \geq 0$ be a martingale with continuous paths. Suppose T is a stopping time with $P(T < \infty) = 1$, and there is a constant K so that $|M_{T \wedge t}| \leq K$ for all t. Then

$$EM_T = EM_0$$

Section 6.3 Martingales, Hitting Times

As in Section 2.4, we can use the linear martingale to compute the:

Example 3.1. Exit distribution from an interval. Define the exit time by $\tau = \min\{t : B_t \notin (a,b)\}$.

(3.3) $$P(B_\tau = b) = \frac{-a}{b-a} \quad \text{and} \quad P(B_\tau = a) = \frac{b}{b-a}$$

Proof. To check τ is a stopping time, note that

$$\{\tau \leq s\}^c = \{\tau > s\} = \{B_r \in (a,b) \text{ for all } r \leq s\}$$

which can be determined by looking at B_r for $r \leq s$. The fact that $P(\tau < \infty) = 1$ can be shown by noting that $\tau \leq T_a$ while $P(T_a < \infty) = 1$ by the formula for the density in (2.5). It is clear that $|B(\tau \wedge t)| \leq |a| + b$, so using (3.2) now we see that

$$0 = EB_\tau = aP(B_\tau = a) + bP(B_\tau = b)$$

Since $P(B_\tau = a) = 1 - P(B_\tau = b)$, we have $0 = a + (b-a)P(B_\tau = b)$. Solving gives the first formula. The second follows from that fact that $P(B_\tau = a) = 1 - P(B_\tau = b)$. □

Using the hitting times $T_x = \min\{t \geq 0 : B_t = x\}$, (3.3) can be written as

$$P(T_a < T_b) = \frac{-a}{b-a} \quad \text{and} \quad P(T_b < T_a) = \frac{b}{b-a}$$

Fixing the value of $a < 0$ and letting $b \to \infty$, it follows that $P(T_a < \infty) = 1$ for $a < 0$. Since this is clearly also true for $a > 0$ and trivial for $a = 0$, we have:

(3.4) Theorem. For all a, $P(T_a < \infty) = 1$.

So Brownian motion will certainly visit every point on the real line. A little thought reveals that it can only do this if and only if

(3.4′) With probability one,

$$\limsup_{t \to \infty} B_t = \infty \quad \text{and} \quad \liminf_{t \to \infty} B_t = -\infty$$

In words, B_t oscillates between ∞ and $-\infty$, and hence visits each value infinitely many times.

For a gambling application of (3.3) consider:

Example 3.2. Optimal doubling in Backgammon. In our idealization, B_t is the probability you will win given the events in the game up to time t. Thus, backgammon is a Brownian motion starting at 1/2 run until it hits 1 (win) or 0 (loss). Initially the "doubling cube" sits in the middle of the board and either player can "double" = tell the other player to play on for twice the stakes or give up and pay the current wager. If a player accepts the double (i.e., decides to play on), she gets possession of the doubling cube and is the only one who can offer the next double.

A doubling strategy is given by two numbers $b < 1/2 < a$, i.e., offer a double when $B_t \geq a$ and give up if the other player doubles and $B_t < b$. It is not hard to see that for the optimal strategy $b^* = 1 - a^*$ and that when $B_t = b^*$ accepting and giving up must have the same payoff. If you accept when your probability of winning is b^* then you lose 2 dollars when your probability hits 0 but you win 2 dollars when your probability of winning hits a^*, since at that moment you can double and the other player gets the same payoff if they give up or play on. If giving up or playing on at b^* is to have the same payoff we must have

$$-1 = \frac{b^*}{a^*} \cdot 2 + \frac{a^* - b^*}{a^*} \cdot (-2)$$

Writing $b^* = c$ and $a^* = 1 - c$ and solving, we have

$$-(1-c) = 2c - 2(1 - 2c) \qquad 1 = 5c$$

so $b^* = 1/5$ and $a^* = 4/5$. In words, you should offer a double when your chances of winning are at least 80% and refuse if your chances of winning are less than 20%. □

b. Quadratic martingale

(3.5) Theorem. $B_t^2 - t$ is a martingale. That is, if $s < t$, then

$$E(B_t^2 - t | B_r, r \leq s) = B_s^2 - s$$

Proof. Again, given B_r, $r \leq s$, the value of B_s is known, while $B_t - B_s$ is independent with mean 0 and variance $t - s$

$$\begin{aligned}E(B_t^2 | B_r, r \leq s) &= E(\{B_s + B_t - B_s\}^2 | B_r, r \leq s) \\ &= B_s^2 + 2B_s E(B_t - B_s | B_r, r \leq s) + E((B_t - B_s)^2 | B_r, r \leq s) \\ &= B_s^2 + 0 + t - s\end{aligned}$$

Subtracting t from each side gives the desired conclusion. □

(3.5) allows us to obtain information about the:

Example 3.3. Mean exit time from an interval. Define the exit time by $\tau = \min\{t : B_t \notin (a,b)\}$.

(3.6) $$E\tau = -ab$$

Why is this true? The martingale $M_t = B_t^2 - t$ does not have the property that $|M_{\tau \wedge t}| \leq K$. However, if we ignore that and apply the stopping theorem, (3.2), and the formula for the distribution of B_τ given in (3.3), then we get

$$0 = E(B_\tau^2 - \tau) = b^2 \cdot \frac{-a}{b-a} + a^2 \cdot \frac{b}{b-a} - E\tau$$

Rearranging now gives

$$E\tau = \frac{-b^2 a + ba^2}{b-a} = -ba \qquad \square$$

Proof. To justify the last computation we let n be a positive integer and consider the stopping time $T = \tau \wedge n$, which has $|M_{T \wedge t}| \leq a^2 + b^2 + n$ for all t. Using (3.2) now as before and rearranging, we have

$$E(\tau \wedge n) = EB_{\tau \wedge n}^2$$

Letting $n \to \infty$ now gives (3.6). $\qquad \square$

To see what (3.6) says we will now derive two consequences. First and simplest is to set $a = -r$, $b = r$ to see that if τ_r is the exit time from $(-r,r)$, then

$$E\tau_r = r^2$$

To see why this should be true note that the scaling relation (1.5) implies that τ_r has the same distribution as $r^2 \tau_1$.

Taking $b = N$ in (3.6) and letting $\tau_{a,N} = \min\{t : B_t \notin (a,N)\}$, we see that if $a < 0$, then

$$ET_a \geq E\tau_{a,N} = -aN \to \infty$$

as $N \to \infty$. The same conclusion holds for $a > 0$ by symmetry, so we have

(3.7)   For all $a \neq 0$, $ET_a = \infty$.

This result does not hold for $a = 0$ since $T_0 = \min\{t \geq 0 : X_t = 0\} = 0$ by definition.

c. Exponential martingale

Before introducing our third and final martingale, we need some algebra,

$$-\frac{x^2}{2u} + \theta x = -\frac{(x-u\theta)^2}{2u} + \frac{u\theta^2}{2}$$

and some calculus.

$$E(\exp(\theta B_u)) = \int e^{\theta x} \frac{1}{\sqrt{2\pi u}} e^{-x^2/2u} \, dx$$

$$= e^{u\theta^2/2} \int \frac{1}{\sqrt{2\pi u}} \exp\left(-\frac{(x-u\theta)^2}{2u}\right) dx = e^{u\theta^2/2}$$

since the integrand is the density of a normal$(u\theta, u)$ distribution.

(3.8) Theorem. $\exp(\theta B_t - t\theta^2/2)$ is a martingale. That is, if $s < t$, then

$$E(\exp(\theta B_t - t\theta^2/2)|B_r, r \leq s) = \exp(\theta B_s - s\theta^2/2)$$

Proof. Again, given B_r, $r \leq s$, the value of B_s is known, while $B_t - B_s$ is an independent normal$(0, t-s)$.

$$E(\exp(\theta B_t)|B_r, r \leq s) = \exp(\theta B_s) E(\exp(\theta(B_t - B_s))|B_r, r \leq s)$$
$$= \exp(\theta B_s + (t-s)\theta^2/2)$$

by our calculus fact with $u = t - s$. Multiplying each side by $\exp(-t\theta^2/2)$ gives the desired result. □

Example 3.4. Probability of ruin when playing a favorable game. Let $X_t = \sigma B_t + \mu t$ be a Brownian motion with drift μ and variance σ^2. Let

$$R_a = \min\{t : X_t = a\} = \min\{t : B_t = (a - \mu t)/\sigma\}$$

(3.9) Theorem. If $\mu > 0$ and $a < 0$, then $P(R_a < \infty) = e^{2\mu a/\sigma^2}$.

To see that this answer is reasonable, note that if a and b are negative, then in order for X_t to reach $a + b$ it must first reach a, an event of probability $P(R_a < \infty)$, and then go from a to $a + b$, an event of probability $P(R_b < \infty)$, so the strong Markov property implies

$$P(R_{a+b} < \infty) = P(R_a < \infty) P(R_b < \infty)$$

This tells us that $P(R_a < \infty) = e^{ca}$ for some c, but cannot tell us that the value of $c = 2\mu/\sigma^2$. To see that the answer should only depend on μ/σ^2 note that
$$X(t/\sigma^2) = \sigma B(t/\sigma^2) + (\mu/\sigma^2)t$$
which by the scaling relation (1.5) has the same distribution as $B_t + (\mu/\sigma^2)t$.

Why is (3.9) true? If we stop the martingale $M_t^\theta = \exp(\theta B_t - t\theta^2/2)$ at time R_a, then $B(R_a) = (a - \mu R_a)/\sigma$, so
$$M^\theta(R_a) = \exp\{\theta(a - \mu R_a)/\sigma - R_a \theta^2/2\} = e^{\theta a/\sigma} \exp\{-R_a(\mu\theta/\sigma + \theta^2/2)\}$$
This suggests choosing θ so that $\mu\theta/\sigma + \theta^2/2 = 0$; i.e., $\theta = -2\mu\sigma$. Taking $T = R_a$ in the optional stopping theorem, (3.2), and supposing there is no contribution from the set $\{R_a = \infty\}$, we have
$$1 = EM^\theta(R_a) = e^{-2\mu a/\sigma^2} P(R_a < \infty)$$
and rearranging gives the desired result. □

Proof of (3.9). Applying the stopping theorem (3.2) at time $R_a \wedge t$ we get

(⋆)
$$\begin{aligned}1 &= E\exp\{-(2\mu/\sigma)B_{R_a \wedge t} - (2\mu/\sigma)^2(R_a \wedge t)/2\} \\ &= E\left(\exp\{-(2\mu/\sigma)B_{R_a} - (2\mu^2/\sigma^2)R_a\}; R_a \leq t\right) \\ &\quad + E\left(\exp\{-(2\mu/\sigma)B_t - (2\mu^2/\sigma^2)t\}; R_a > t\right)\end{aligned}$$

Since $B(R_a) = (a - \mu R_a)/\sigma$, the second line is
$$E\left(e^{-2\mu a/\sigma^2}; R_a \leq t\right) = e^{-2\mu a/\sigma^2} P(R_a \leq t) \to e^{-2\mu a/\sigma^2} P(R_a < \infty)$$
as $t \to \infty$. To deal with the third line in (⋆), we note that the strong law of large numbers in Exercise 1.6 implies that with probability one $B_t/t \to 0$. Thus on the event $\{R_a > t\}$ we have
$$\exp(-(2\mu/\sigma)B_t - (2\mu^2/\sigma^2)t) \to 0$$
and the contribution to the expected value from $R_a > t$ tends to 0. Combining the last three displays gives the desired result. □

d. Higher-order martingales

The result in (3.8) says
$$E(\exp(\theta B_t - \theta^2 t/2)|B_r, r \leq s) = \exp(\theta B_s - \theta^2 s/2)$$

Differentiating with respect to θ (and not worrying about the details) gives

(*) $\quad E((B_t - \theta t)\exp(\theta B_t - \theta^2 t/2)|B_r, r \le s) = (B_s - \theta s)\exp(\theta B_s - \theta^2 s/2)$

Setting $\theta = 0$, we get the martingale in (3.1):

$$E(B_t|B_r, r \le s) = B_s$$

Differentiating (*) again gives

(**) $\quad \begin{aligned} E(\{(B_t - \theta t)^2 - t\}\exp(\theta B_t - \theta^2 t/2)|B_r, r \le s) \\ = \{(B_s - \theta s)^2 - s\}\exp(\theta B_s - \theta^2 s/2) \end{aligned}$

Setting $\theta = 0$, we get the martingale in (3.5):

$$E(B_t^2 - t|B_r, r \le s) = B_s^2 - s$$

Having reinvented our linear and quadratic martingales, it is natural to differentiate more to find a new martingales. To get organized for doing this, we let

$$f(\theta, x, t) = \exp(\theta x - \theta^2 t/2)$$

let $f_k(\theta, x, t)$ be the k the derivative w.r.t. θ, and let $h_k(x, t) = f_k(0, x, t)$. Repeating the arguments for (*) and (**) we see that:

(3.10) **Theorem.** $h_k(B_t, t)$ is a martingale.

As noted above, we have already seen $h_1(B_t, t)$ and $h_2(B_t, t)$, but the other ones are new:

k	$f_k(\theta, x, t)$	$h_k(B_t, t)$
1	$(x - \theta t)f(\theta, x, t)$	B_t
2	$\{(x - \theta t)^2 - t\}f(\theta, x, t)$	$B_t^2 - t$
3	$\{(x - \theta t)^3 - 3t(x - \theta t)\}f(\theta, x, t)$	$B_t^3 - 3tB_t$
4	$\{(x - \theta t)^4 - 6t(x - \theta t)^2 + 3t^2\}f(\theta, x, t)$	$B_t^4 - 6tB_t^2 + 3t^2$

The next result illustrates the use of the martingale $h_4(B_t, t)$:

(3.11) If $\tau_a = \inf\{t : B_t \notin (-a, a)\}$, then

$$E\tau_a^2 = 5a^4/3$$

To see why the answer is a multiple of a^4, note that the scaling relationship tells us that τ_a has the same distribution as $a^2\tau_1$, so $E\tau_a^2 = a^4 E\tau_1^2$.

Proof. Using (3.2) at time $\tau_a \wedge t$, we have

$$E(B(\tau_a \wedge t)^4 - 6(\tau_a \wedge t)B(\tau_a \wedge t)^2) = -3E(\tau_a \wedge t)^2$$

From (3.6) we know that $E\tau_a = a^2 < \infty$. Letting $t \to \infty$, gives

$$a^4 - 6a^2 E\tau_a = -3E(\tau_a^2)$$

Plugging in $E\tau_a = a^2$ and doing a little arithmetic gives the desired result. □

6.4. Option Pricing in Discrete Time

In the next section we will use our knowledge of Brownian motion to derive the Black–Scholes formula. To prepare for that computation, we will consider the much simpler problem of pricing options when there are a finite number of time periods and two possible outcomes at each stage. The restriction to two outcomes is not as bad as one might think. One justification for this is that we are looking at the process on a very slow time scale, so at most one interesting event happens (or not) per time period. We begin by considering a very simple special case.

Example 4.1. Two-period binary tree. Suppose that a stock price starts at 100 at time 0. At time 1 (one day or one month or one year later) it will either be worth 120 or 90. If the stock is worth 120 at time 1, then it might be worth 130 or 110 at time 2. If the price is 90 at time 1, then the possibilities at time 2 are 120 and 80. The last three sentences can be simply summarized by the following tree.

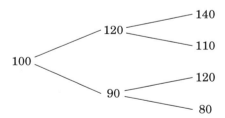

Suppose now that you are offered a **European call option** with **strike price** 100 and **expiry** 2. This means you have an option to buy the stock (but not an obligation to do so) for 100 at time 2, i.e., after seeing the outcome of the first and second stages. If the stock price is 80, you will not exercise the option to purchase the stock and your profit will be 0. In the other cases you will choose to buy the stock at 100 and then immediately sell it at X_2 to get a payoff of $X_2 - 100$ where X_2 is the stock price at time 2. Combining the two cases we can write the payoff in general as $(X_2 - 100)^+$, where $z^+ = \max\{z, 0\}$ denotes the positive part of z.

Our problem is to figure out what is the right price for this option. At first glance this may seem impossible since we have not assigned probabilities to the various events. However, it is a miracle of "pricing by the absence of arbitrage" that in this case we do not have to assign probabilities to the events to compute the price. To explain this we start by considering a small piece of the tree. When $X_1 = 90$, X_2 will be 120 ("up") or 80 ("down") for a profit of 30 or a loss of 10, respectively. If we pay c for the option, then when X_2 is up we make a profit of $20 - c$, but when it is down we make $-c$. The last two sentences are summarized in the following table

	stock	option
up	30	$20 - c$
down	-10	$-c$

Suppose we buy x units of the stock and y units of the option, where negative numbers indicate that we sold instead of bought. One possible strategy is to choose x and y so that the outcome is the same if the stock goes up or down:

$$30x + (20 - c)y = -10x + (-c)y$$

Solving, we have $40x + 20y = 0$ or $y = -2x$. Plugging this choice of y into the last equation shows that our profit will be $(-10 + 2c)x$. If $c > 5$, then we can make a large profit with no risk by buying large amounts of the stock and selling twice as many options. Of course, if $c < 5$, we can make a large profit by doing the reverse. Thus, in this case the only sensible price for the option is 5.

A scheme that makes money without any possibility of a loss is called an **arbitrage opportunity**. It is reasonable to think that these will not exist in financial markets (or at least be short-lived) since if and when they exist people take advantage of them and the opportunity goes away. Using our new terminology we can say that the only price for the option which is consistent with absence of arbitrage is $c = 5$, so that must be the price of the option (at time 1 when $X_1 = 90$).

Before we try to tackle the whole tree to figure out the price of the option at time 0, it is useful to look at things in a different way. Generalizing our example, let $a_{i,j}$ be the profit for the ith security when the jth outcome occurs.

(4.1) Theorem. *Exactly one of the following holds:*

(i) *There is a betting scheme x_i so that $\sum_{i=1}^m x_i a_{i,j} \geq 0$ for each j and $\sum_{i=1}^m x_i a_{i,k} > 0$ for some k.*

Section 6.4 Option Pricing in Discrete Time

(ii) There is a probability vector $p_j > 0$ so that $\sum_{j=1}^n a_{i,j} p_j = 0$ for all i.

Here an x satisfying (i) is an arbitrage opportunity. We never lose any money but for at least one outcome we gain a positive amount. Turning to (ii), the vector p_j is called a martingale measure since if the probability of the jth outcome is p_j, then the expected change in the price of the ith stock is equal to 0. Combining the two interpretations we can restate (4.1) as:

(4.2) Theorem. *There is no arbitrage if and only if there is a strictly positive probability vector so that all the stock prices are martingale.*

Why is (4.1) true? One direction is easy. If (i) is true, then for any strictly positive probability vector $\sum_{i=1}^m \sum_{j=1}^n x_i a_{i,j} p_j > 0$, so (ii) is false.

Suppose now that (i) is false. The linear combinations $\sum_{i=1}^m x_i a_{i,j}$ when viewed as vectors indexed by j form a linear subspace of n-dimensional Euclidean space. Call it $\mathcal{L}$. If (i) is false, this subspace intersects the positive orthant $\mathcal{O} = \{y : y_j \geq 0 \text{ for all } j\}$ only at the origin. By linear algebra we know that $\mathcal{L}$ can be extended to an $n-1$ dimensional subspace $\mathcal{H}$ that only intersects $\mathcal{O}$ at the origin.

Since $\mathcal{H}$ has dimension $n-1$, it can be written as $\mathcal{H} = \{y : \sum_{j=1}^n y_j p_j = 0\}$. Since for each fixed i the vector $a_{i,j}$ is in $\mathcal{L} \subset \mathcal{H}$, (ii) holds. To see that all the $p_j > 0$ we leave it to the reader to check that if not, there would be a non-zero vector in $\mathcal{O}$ that would be in $\mathcal{H}$. □

To apply (4.1) to our simplified example we begin by noting that in this case $a_{i,j}$ is given by

		$j = 1$	$j = 2$
stock	$i = 1$	30	-10
option	$i = 2$	$20 - c$	$-c$

By (4.2) if there is no arbitrage, then there must be an assignment of probabilities p_j so that

$$30p_1 - 10p_2 = 0 \qquad (20 - c)p_1 + (-c)p_2 = 0$$

From the first equation we conclude that $p_1 = 1/4$ and $p_2 = 3/4$. Rewriting the second we have

$$c = 20p_1 = 20 \cdot (1/4) = 5$$

To generalize from the last calculation to finish our example we note that the equation $30p_1 - 10p_2 = 0$ says that under p_j the stock price is a martingale (i.e., the average value of the change in price is 0), while $c = 20p_1 + 0p_2$ says that the price of the option is then the expected value under the martingale probabilities. Using these ideas we can quickly complete the computations in our example. When $X_1 = 120$ the two possible scenarios lead to a change of $+20$ or -5, so the relative probabilities of these two events should be $1/5$ and $4/5$. When $X_0 = 0$ the possible price changes on the first step are $+20$ and -10, so their relative probabilities are $1/3$ and $2/3$. Making a table of the possibilities, we have

X_1	X_2	probability	$(X_2 - 100)^+$
120	140	$(1/3) \cdot (1/5)$	40
120	115	$(1/3) \cdot (4/5)$	15
90	120	$(2/3) \cdot (1/4)$	20
90	80	$(2/3) \cdot (3/4)$	0

so the value of the option is

$$\frac{1}{15} \cdot 40 + \frac{4}{15} \cdot 15 + \frac{1}{6} \cdot 20 = \frac{80 + 120 + 100}{30} = 10 \qquad \square$$

The last derivation may seem a little devious, so we will now give a second derivation of the price of the option. In the scenario described above, our investor has four possible actions:

A_0. Put \$1 in the bank and end up with \$1 in all possible scenarios.

A_1. Buy one share of stock at time 0 and sell it at time 1.

A_2. Buy one share at time 1 if the stock is at 120, and sell it at time 2.

A_3. Buy one share at time 1 if the stock is at 90, and sell it at time 2.

These actions produce the following payoffs in the indicated outcomes

X_1	X_2	A_0	A_1	A_2	A_3	option
120	140	1	20	20	0	40
120	115	1	20	-5	0	15
90	120	1	-10	0	30	20
90	80	1	-10	0	-10	0

Noting that the payoffs from the four actions are themselves vectors in four-dimensional space, it is natural to think that by using a linear combination of these actions we can reproduce the option exactly. To find the coefficients we write four equations in four unknowns,

$$z_0 + 20z_1 + 20z_2 = 40$$
$$z_0 + 20z_1 - 5z_2 = 15$$
$$z_0 - 10z_1 + 30z_3 = 20$$
$$z_0 - 10z_1 - 10z_3 = 0$$

Subtracting the second equation from the first and the fourth from the third gives $25z_2 = 25$ and $40z_3 = 20$ so $z_2 = 1$ and $z_3 = 1/2$. Pugging in these values, we have two equations in two unknowns:

$$z_0 + 20z_1 = 20 \qquad z_0 - 10z_1 = 5$$

Taking differences, we conclude $30z_1 = 15$, so $z_1 = 1/2$ and $z_0 = 10$.

The reader may have already noticed that $z_0 = 10$ is the option price. This is no accident. What we have shown is that with \$10 cash we can buy and sell shares of stock to produce the outcome of the option in all cases. In the terminology of Wall Street, $z_1 = 1/2$, $z_2 = 1$, $z_3 = 1/2$ is a **hedging strategy** that allows us to **replicate the option**. Once we can do this it follows that the fair price must be \$10. To do this note that if we could sell it for \$12 then we can take \$10 of the cash to replicate the option and have a sure profit of \$2.

6.5. The Black–Scholes Formula

In this section we will find the fair price of a European call option $(X_t - K)^+$ with strike price K and expiry t based on a stock price that follows

(5.1) $$X_t = X_0 \cdot \exp(\mu t + \sigma B_t)$$

Here B_t is a standard Brownian motion, μ is the exponential growth rate of the stock, and σ is its volatility.

In writing the model we have assumed that the growth rate and volatility of the stock are constant. If we also assume that the interest rate r is constant, then the discounted stock price is

$$e^{-rt} X_t = X_0 \cdot \exp((\mu - r)t + \sigma B_t)$$

Here we have to multiply by e^{-rt}, since \$1 at time t has the same value as e^{-rt} dollars today.

Extrapolating wildly from (4.1), we can say that any consistent set of prices must come from a martingale measure. Combining this with the fact proved in (3.8) that $\exp(\theta B_t - (t\theta^2/2))$ is martingale, we see that if

(5.2) $$\mu = r - \sigma^2/2$$

then the discounted stock price, $e^{-rt} X_t$ is a martingale. To compute the value of the call option, we need to compute its value in the model in (5.1) for this special

262   Chapter 6   Brownian Motion

value of μ. Using the fact that $\log(X_t/X_0)$ has a normal$(\mu t, \sigma^2 t)$ distribution, we see that

$$E(e^{-rt}(X_t - K)^+) = e^{-rt} \int_{\log(K/X_0)}^{\infty} (X_0 e^y - K) \frac{1}{\sqrt{2\pi\sigma^2 t}} e^{-(y-\mu t)^2/2\sigma^2 t} \, dy$$

Changing variables $y = \mu t + w\sigma\sqrt{t}$, $dy = \sigma\sqrt{t}\, dw$ the integral is equal to

$$(\star) \qquad = e^{-rt} X_0 e^{\mu t} \frac{1}{\sqrt{2\pi}} \int_{\alpha}^{\infty} e^{w\sigma\sqrt{t}} e^{-w^2/2} \, dw - e^{-rt} K \frac{1}{\sqrt{2\pi}} \int_{\alpha}^{\infty} e^{-w^2/2} \, dw$$

where $\alpha = (\log(K/X_0) - \mu t)/\sigma\sqrt{t}$. The first integral

$$\frac{1}{\sqrt{2\pi}} \int_{\alpha}^{\infty} e^{w\sigma\sqrt{t}} e^{-w^2/2} \, dw = e^{t\sigma^2/2} \int_{\alpha}^{\infty} \frac{1}{\sqrt{2\pi}} e^{-(w-\sigma\sqrt{t})^2/2} \, dw$$
$$= e^{t\sigma^2/2} \, P(\text{normal}(\sigma\sqrt{t}, 1) > \alpha)$$

The last probability can be written in terms of the distribution function Φ of a standard normal χ, i.e., $\Phi(t) = P(\chi \le t)$, by noting

$$P(\text{normal}(\sigma\sqrt{t}, 1) > \alpha) = P(\chi > \alpha - \sigma\sqrt{t})$$
$$= P(\chi \le \sigma\sqrt{t} - \alpha) = \Phi(\sigma\sqrt{t} - \alpha)$$

where in the middle equality we have used the fact that χ and $-\chi$ have the same distribution. Using the last two computations in $(\star)$ converts it to

$$e^{-rt} X_0 e^{\mu t} e^{\sigma^2 t/2} \Phi(\sigma\sqrt{t} - \alpha) - e^{-rt} K \Phi(-\alpha)$$

Using (5.2) now the expression simplifies to the

(5.3) Black–Scholes formula. *The price of the European call option* $(X_T - K)^+$ *is given by*

$$X_0 \Phi(\sigma\sqrt{t} - \alpha) - e^{-rt} K \Phi(-\alpha)$$

where $\alpha = \{\log(K/X_0 e^{\mu t})\}/\sigma\sqrt{t}$ *and* $\mu = r - \sigma^2/2$.

To try to come to grips with this ugly formula note that $K/X_0 e^{\mu t}$ is the ratio of the strike price to the expected value of the stock at time t under the martingale probabilities, while $\sigma\sqrt{t}$ is the standard deviation of $\log(X_t/X_0)$.

Example 5.1. Microsoft call options. The February 23, 1998, *Wall Street Journal* listed the following prices for July call options on Microsoft stock.

Section 6.5 The Black–Scholes Formula

strike	75	80	85
price	11	8 1/8	5 1/2

On this date Microsoft stock was trading at 81 5/8, while the annual interest rate was about 4% per year. Should you buy the call option with strike 80?

Solution. The answer to this question will depend on your opinion of the volatility of the market over the period. Suppose that we follow a traditional rule of thumb and decide that $\sigma = 0.3$; i.e., over a one-year period a stock's price might change by about 30% of its current value. In this case the drift rate for the martingale measure is

$$\mu = r - \sigma^2/2 = .04 - (.09)/2 = .04 - .045 = -.005$$

and so the log ratio is

$$\log(K/X_0 e^{\mu t}) = \log(80/(81.625 e^{-.005(5/12)})) = \log(80/81.455) = -.018026$$

Five months corresponds to $t = 5/12$, so the standard deviation

$$\sigma\sqrt{t} = .3\sqrt{5/12} = .19364$$

and $\alpha = -.018026/.19364 = -.09309$. Plugging in now, we have a price of

$$81.625\Phi(.19365 + .09309) - e^{-.04(5/12)} 80\Phi(.09309)$$
$$= 81.625\Phi(.28674) - 78.678\Phi(.09309)$$
$$= 81.625(.6128) - 78.678(.5371) = 50.02 - 42.25 = 7.76$$

This is somewhat lower than the price quoted in the paper. There are two reasons for this. First, the options listed in the *Wall Street Journal* are **American call options**. The holder has the right to exercise at any time during the life of the option. Since one can ignore the additional freedom to exercise early, American options are at least as valuable as their European counterparts.

Second, and perhaps more importantly, we have not spent much effort on our estimate of r and σ. In connection with the last point it is interesting to note that:

(5.4) Lemma. *If we fix the value of r, then the Black–Scholes price is an increasing function of the volatility.*

Thus for any given interest rate, we can, by varying the volatility, match the Black–Scholes price exactly. The magic value of the volatility that gives us the option price is called the **implied volatility**. A consequence of this observation is that we cannot test the adequacy of the Black–Scholes model by looking only at one price.

Example 5.2. Intel call options. Again consulting the *Wall Street Journal* for February 23, 1998, we find the following prices listed for July call options on Intel stock, which was trading at 94 3/16.

strike	70	75	80	85	90	95	100	105
price	26	22	18	$14\frac{1}{2}$	$11\frac{3}{8}$	$8\frac{3}{4}$	$6\frac{1}{2}$	$4\frac{3}{8}$

We leave it as an exercise for the reader to compute the volatilities implied these prices. I have not done this, but folk wisdom suggests that the volatilities will not be constant but "smile" in some convex sort of way. My personal intuition is that large movements in the market are more likely than what is predicted by the Brownian model, so options that are well "out of the money" are overpriced. This situation is also reasonable from an economic point of view as well. Such financial instruments have low payoff relative to their uncertainty and hence are not attractive to sell.

Replication. By analogy with the discrete case, we might ask: Can the option be replicated by using cash and trading in the underlying security? The answer to this questions is yes. Its proof is beyond the scope of this book. However, it is possible to describe the hedging strategy. Let

$$v(s, x) = E((X_t - K)^+ | X_s = x)$$

be the value of the option at time s when the stock price is x where the expected value is computed with respect to the martingale measure. A result known as Ito's formula allows us to conclude

$$v(t, X_t) = v(0, X_0) + \int_0^t \frac{\partial v}{\partial x}(s, X_s) \, dX_s$$

The term $v(0, X_0)$ is the cash we put in to start the replication process. The second term is a "stochastic integral." Intuitively, at time s we hold $(\partial v/\partial x)(s, X_s)$ of the stock, and the second term computes our profit (or loss from using this trading strategy. At time t, $v(t, X_t) = (X_t - K)^+$ is the outcome of the option, so this strategy reproduces the option exactly. As in Section 6.4, the assumption of no arbitrage implies that the price of the option must be $v(0, X_0)$

6.6. Exercises

6.1. Show that if $s < t$, the conditional distribution $(B_s|B_t = z)$ is normal with mean zs/t and variance $s(t-s)/t$.

6.2. (a) Compute the joint distribution $((B_r, B_s)|B_t = 0)$ to show that the covariance is $r(t-s)/t$. (b) Conclude that $B_s^{0,t} = B_s - (s/t)B_t$, $0 \le s \le t$ has the same distribution as $(\{B_s, 0 \le s \le t\}|B_t = 0)$.

6.3. (a) Show that the joint distribution $((B_r - rz/t, B_s - sz/t)|B_t = z)$ is the same as that of $((B_r, B_s)|B_0 = 0)$ (b) Use the result of the previous exercise to conclude that $((B_r, B_s)|B_t = z)$ is a bivariate normal in which the components have means $\mu_r = zr/t$, $\mu_s = zs/t$, variances $\sigma_r^2 = r(t-r)/t$, $\sigma_s^2 = s(t-s)/t$, and covariance $E((B_r, B_s)|B_t = z) = r(t-s)/t$. (b) Show that $B_s^{z,t} = B_s + (s/t)(z - B_t)$, $0 \le s \le t$ has the same distribution as $(\{B_s, 0 \le s \le t\}|B_t = z)$.

6.4. Let $Y_t = \int_0^t B_s\,ds$. Find (a) EY_t, (b) EY_t^2, (c) $E(Y_s, Y_t)$.

6.5. Find the conditional distribution of $Y_t = \int_0^t B_s\,ds$ given $B_t = x$.

6.6. *Strong law.* (a) Use the strong law of large numbers to show that as $n \to \infty$ through the integers, we have $B(n)/n \to 0$. (b) Use (ii) of Exercise 2.1 in Chapter 5 to extend the previous conclusion to show that as $u \to \infty$ through the real numbers, we have $B(u)/u \to 0$.

6.7. *Inversion.* Let B_t be a standard Brownian motion. Define a process X by setting $X_0 = 0$ and $X_t = tB(1/t)$ for $t > 0$. Show that X_t is a Gaussian process with $EX_t = 0$ and $E(X_s X_t) = s \wedge t$, so X_t is a standard Brownian motion. (b) Let $u = 1/t$ in the previous exercise to conclude that $\lim_{t \to 0} X_t = 0$.

6.8. Let $B_t^0 = B_t - tB_1$ be Brownian bridge and let $X_t = (1+t)B^0(t/(1+t))$. Show that X_t is a standard Brownian motion.

6.9. Check that $(1 - 3y^{-4})e^{-y^2/2} \le e^{-y^2/2} \le e^{-x^2/2}e^{-x(y-x)}$ for $y \ge x$. Then integrate to conclude

$$(x^{-1} - x^{-3})\exp(-x^2/2) \le \int_x^\infty \exp(-y^2/2)\,dy \le x^{-1}\exp(-x^2/2)$$

6.10. Show that if X and Y are independent normal$(0,\sigma^2)$, then $U = (X+Y)/2$ and $V = (X-Y)/2$ are independent normal$(0,\sigma^2/2)$.

6.11. *Levy's interpolation construction of Brownian motion.* We construct the Brownian motion only for $0 \le t \le 1$. For each $n \ge 0$ let $Y_{n,1}, \ldots, Y_{n,2^n}$ be independent normal(0,1). Start by setting $B(0) = 0$ and $B(1) = Y_{0,1}$. Assume now that $B(m/2^n)$ has been defined for $0 \le m \le 2^n$ and let

$$B((2m+1)/2^{n+1}) = \frac{\{B((m+1)/2^n) - B(m/2^n)\} + 2^{-n/2}Y_{n,m+1}}{2}$$

(a) Use the previous exercise to show that the increments

$$B(k/2^{n+1}) - B((k-1)/2^{n+1}) \quad 1 \le k \le 2^{n+1}$$

are independent normal$(0, 2^{-(n+1)})$. (b) Let $B_n(m/2^n) = B(m/2^n)$ and let $B_n(t)$ be linear on each interval $[m/2^n, (m+1)/2^n]$. Show that

$$\max_t |B_n(t) - B_{n+1}(t)| = \frac{1}{2} \cdot 2^{-n/2} \max_m |Y_{n+1,m}|$$

(c) Use Exercise 6.9 to conclude that

$$P\left(\max_{1 \le m \le 2^n} |Y_{n+1,m}| \ge \sqrt{2n}\right) \le 2 \cdot 2^n \cdot e^{-n}$$

(d) Use (b) and (c) to conclude that there is a limit $B(t)$ so that for large n:

$$\max_t |B_n(t) - B(t)| \le \sum_{k=n}^{\infty} 2^{-k/2}\sqrt{2k}$$

This shows that $B(t)$, $0 \le t \le 1$ is a uniform limit of continuous functions. It then follows from a result in real analysis that $B(t)$ is continuous.

6.12. *Nondifferentiability of Browian paths.* The point of this exercise is to show that with probability one, Brownian paths are not Lipschitz continuous (and hence not differentiable) at any point. Fix a constant $C < \infty$ and let $A_n = \{\omega : \text{there is an } s \in [0, 1] \text{ so that } |B_t - B_s| \le C|t - s| \text{ when } |t - s| \le 3/n\}$. For $1 \le k \le n-2$ let

$$Y_{k,n} = \max\left\{\left|B\left(\frac{k+j}{n}\right) - B\left(\frac{k+j-1}{n}\right)\right| : j = 0, 1, 2\right\}$$

$G_n = \{ \text{at least one } Y_{k,n} \le 5C/n\}$

(a) Show that $A_n \subset G_n$. (b) Show that $P(G_n) \to 0$. (c) Conclude that $P(A_m) = 0$ for all m, which gives the desired conclusion.

6.13. *Quadratic variation.* Fix t and let $\Delta_{m,n} = B(tm2^{n-1}) - B(t(m-1)2^{n-1})$. Compute

$$E\left(\sum_{m \le 2^n} \Delta_{m,n}^2 - t\right)^2$$

and conclude that $\sum_{m\leq 2^n} \Delta^2_{m,n} \to t$ as $n \to \infty$.

6.14. Let $\varphi_a(\lambda) = E\exp(-\lambda T_a)$ be the Laplace transform of T_a. (a) Use (2.3) to conclude that $\varphi_a(\lambda)\varphi_b(\lambda) = \varphi_{a+b}(\lambda)$. (b) Use the scaling relationship to conclude that T_a has the same distribution as $a^2 T_1$. (c) Combine the first two parts to conclude that there is a constant κ so that $\varphi_a(\lambda) = \exp(-a\kappa\sqrt{\lambda})$.

6.15. Let (B_t^1, B_t^2) be a two dimensional Brownian motion starting from 0. Let $T_a^1 = \min\{t : B_t^1 = a\}$ and let $Y_a = B^2(T_a^1)$. (a) Use the strong Markov property to conclude that if $a < b$ then Y_a and $Y_b - Y_a$ are independent. (b) Use the scaling relationship (1.5) to conclude that Y_a has the same distribution as aY_1. (c) Use (b) the fact that Y_a and $-Y_a$ have the same distribution to show that the Fourier transform $E\exp(i\theta Y_a) = \exp(-a\kappa|\theta|)$ for some constant κ.

6.16. Continuing with the set-up of the previous problem, use the density function of T_a given in (2.5) to show that Y_a has a Cauchy distribution.

6.17. Let $x > 0$. Compute $P(B_t > y, \min_{0\leq u\leq t} B_t > 0 | B_0 = x)$ and then differentiate to find the density function.

6.18. Let $M_t = \max_{0\leq u\leq t} B_u$. Find $P(M_t \geq x, B_t \leq y)$ for $x > y$ then differentiate to find the joint density of (M_t, B_t).

6.19. Use the fact that $A_s = B_t - B_{t-s}$ is a Brownian motion and the previous exercise to conclude

$$P(B_t > x | B_u \geq 0 \text{ for all } 0 \leq u \leq t) = P(M_t > x | B_t = M_t) = e^{-x^2/2t}$$

6.20. Let $r < s < t$. Find the conditional probability that a standard Brownian motion is not 0 in the interval (s,t) given that it was not 0 in the interval (r,s).

6.21. Let $R_t = \min\{s > t : B_s = 0\}$. Find the density function $P_0(R_t = u)$.

6.22. Let $T_a = \inf\{t : B_t = a\}$. Compute $P(T_1 < T_{-2} < T_3)$.

6.23. Show that for each $a > 1$ there is a unique b so that $P(T_1 < T_{-a} < T_b) = 1/2$. Find the value of b for $a = 2$.

6.24. Let $(B_t^1, \ldots B_t^d)$ be a d-dimensional Brownian motion and let

$$R_t = \sqrt{(B_t^1)^2 + \cdots + (B_t^d)^2}$$

be its distance from the origin, and $S_r = \min\{t : R_t = r\}$ be the first time the distance from the origin is equal to r. (a) Show that $R_t^2 - td$ is a martingale. (b) Imitate the proof of (3.6) to conclude that $ES_r = r^2/d$.

6.25. (a) Integrate by parts to show that

$$\int x^m e^{-x^2/2}\, dx = (m-1)\int x^{m-2} e^{-x^2/2}\, dx$$

(b) Use this result to conclude that if χ has a standard normal distribution then for any integer $n \geq 1$

$$E\chi^{2n} = (2n-1)(2n-3)\cdots 3\cdot 1 \quad \text{and} \quad E\chi^{2n-1} = 0$$

That is, $E\chi^4 = 3$, $E\chi^6 = 15$, etc., while all odd moments are 0.

6.26. Verify directly that

$$E(B_t^4 - 6tB_t^2 + 3t^2 | B_r, r \leq s) = B_s^4 - 6sB_s^2 + 3s^2$$

6.27. Let $T_a = \min\{t : B_t = a\}$. Use the exponential martingale in part (c) of Section 6.3 to conclude that $E\exp(-\lambda T_a) = e^{-a\sqrt{2\lambda}}$.

6.28. Let $\tau = \inf\{t : B_t \notin (a,b)\}$ and let $\lambda > 0$. Use the strong Markov property to show

$$E\exp(-\lambda T_a) = E(e^{-\lambda \tau}; T_a < T_b) + E(e^{-\lambda \tau}; T_b < T_a)E\exp(-\lambda T_{a-b})$$

(ii) Interchange the roles of a and b to get a second equation, use the previous exercise, and solve to get

$$E_x(e^{-\lambda T}; T_a < T_b) = \sinh(\sqrt{2\lambda}b)/\sinh(\sqrt{2\lambda}(b-a))$$
$$E_x(e^{-\lambda T}; T_b < T_a) = \sinh(-\sqrt{2\lambda}a)/\sinh(\sqrt{2\lambda}(b-a))$$

6.29. As in Example 3.4, let $X_t = \sigma B_t + \mu t$ with $\mu > 0$ and $R_a = \min\{t : X_t = a\}$ where $a < 0$. (i) Set $\mu\theta/\sigma + \theta^2/2 = \lambda$ and solve to get a negative solution $\theta = -\nu - (\nu^2 + 2\lambda)^{1/2}$ where $\nu = \mu/\sigma$. (ii) Use the reasoning for (3.9) to show

$$E\exp(-\lambda R_a) = \exp(-\theta a/\sigma) = \exp\left(\frac{a}{\sigma}\{\nu + (\nu^2 + 2\lambda)^{1/2}\}\right)$$

Taking $\lambda = 0$, we get (3.9): $P(R_a < \infty) = \exp(2\mu a/\sigma^2)$.

6.30. Again let $X_t = \sigma B_t + \mu t$ where $\mu > 0$ but now let $R_b = \min\{t : X_t = b\}$ where $b > 0$. Show that $ER_b = b/\mu$.

6.31. As in the previous exercise, let $X_t = \sigma B_t + \mu t$ with $\mu > 0$ and $R_b = \min\{t : X_t = b\}$ where $b > 0$. (i) Set $\mu\theta/\sigma + \theta^2/2 = \lambda$ and solve to get a positive solution $\theta = -\nu + (\nu^2 + 2\lambda)^{1/2}$ where $\nu = \mu/\sigma$. (ii) Use the reasoning for (3.9) to show

$$E\exp(-\lambda R_b) = \exp(-\theta b/\sigma) = \exp\left(\frac{b}{\sigma}\{\nu - (\nu^2 + 2\lambda)^{1/2}\}\right)$$

Taking $\lambda = 0$ this time we have $P(R_b < \infty) = 1$.

6.32. Let $\tau_a = \inf\{t : B_t \notin (-a,a)\}$ and recall $\cosh(x) = (e^x + e^{-x})/2$. Show that $\exp(-\theta^2 t/2)\cosh(\theta B_t)$ is a martingale and use this to conclude that

$$E\exp(-\lambda\tau_a) = 1/\cosh(a\sqrt{2\lambda})$$

6.33. Let $p_t(x,y) = (2\pi t)^{-1/2} e^{-(y-x)^2/2t}$. Check that

$$\frac{\partial}{\partial t} p_t(x,y) = \frac{1}{2}\frac{\partial^2}{\partial y^2} p_t(x,y)$$

6.34. Let $u(t,x)$ be a function that satisfies

(*) $\quad \dfrac{\partial u}{\partial t} + \dfrac{1}{2}\dfrac{\partial^2 u}{\partial x^2} = 0 \quad$ and $\quad \left|\dfrac{\partial^2 u}{\partial x^2}(t,x)\right| \le C_T \exp(x^2/(t+\epsilon)) \quad$ for $t \le T$

Show that $u(t, B_t)$ is a martingale by using the previous exercise to conclude

$$\frac{\partial}{\partial t} Eu(t, B_t) = \int \frac{\partial}{\partial t}(p_t(0,y)u(t,y))\,dy = 0$$

Examples of functions that satisfy (*) are $\exp(\theta x - \theta^2 t/2)$, x, $x^2 - t$, ...

6.35. Find a martingale of the form $B_t^6 - at B_t^4 + bt^2 B_t^2 - ct^3$ and use it to compute the third moment of $\tau_a = \inf\{t : B_t \notin (-r,r)\}$. Note that scaling implies $E\tau_a = ca^6$.

6.36. (a) Use Exercise 6.34 to show that $(1+t)^{-1/2}\exp(B_t^2/2(1+t))$ is a martingale. (b) Use (a) to conclude that for large integers n, we have

$$|B_n| < \sqrt{(3.002)n \log n}$$

6.37. *Law of the iterated logarithm.* By working much harder than in the previous exercise one can show that

$$\limsup_{t\to\infty} B_t/\sqrt{2t\log\log t} = 1$$

Use the inversion $X(t) = tB(1/t)$ to conclude that

$$\limsup_{t\to 0} B_t/\sqrt{2t\log\log(1/t)} = 1$$

6.38. The Cornell hockey team is playing a game against Harvard: it will either win lose or draw. A gambler offers you the following three payoffs, each for a $1 bet.

	win	lose	draw
Bet 1	0	1	1.5
Bet 2	2	2	0
Bet 3	.5	1.5	0

Assume you are able to buy any amounts (even negative) of these bets. Is there an arbitrage opportunity?

6.39. Suppose Microsoft stock sells for 100 while Netscape sells for 50. Three possible outcomes of a court case will have the following impact on the two stocks.

	Microsoft	Netscape
1 (win)	120	30
2 (draw)	110	55
3 (lose)	84	60

What should we be willing to pay for an option to buy Netscape for 50 after the court case is over?

6.40. *Put-call parity.* $(K - X_T)^+$ is a European put with strike price K and expiry T. Show that the put price can be obtained from the call price by

$$Ee^{-rt}(K - X_T)^+ = Ee^{-rt}(X_T - K)^+ - X_0$$

References

Consult the following sources to learn more about the subject.

Athreya, K.B. and Ney, P.E. (1972) *Branching Processes.* Springer-Verlag, New York

Bailey, N.T.J. (1964) *The Elements of Stochastic Processes: With Applications to the Natural Sciences.* John Wiley and Sons.

Barbour, A.D., Holst, L., and Janson, S. (1992) *Poisson Approximation.* Oxford U. Press

Bhattacharya, R.N. and Waymire, C. *Stochastic Processes with Applications.* John Wiley and Sons, New York

Chung, K.L. (1967) *Markov Chains with Stationary Transition Probabilities.* Second edition, Springer-Verlag, New York

Cox, D.R. and Miller, H.D. (1965) *The Theory of Stochastic Processes.* Methuen & Co, Ltd., London

Doob, J.L. (1953) *Stochastic Processes.* John Wiley and Sons, New York

Durrett, R. (1993) *Essentials of Probability.* Duxbury Press, Belmont, CA

Durrett, R. (1995) *Probability: Theory and Examples.* Duxbury Press, Belmont, CA

Durrett, R. (1996) *Stochastic Calculus: A Practical Introduction.* CRC Press, Boca Raton, FL

Feller, W. (1968) *An Introduction to Probability Theory and its Applications.* Third edition, John Wiley and Sons, New York

Hoel, P.G., Port, S.C. and Stone, C.J. (1972) *Introduction to Stochastic Processes.* Houghton-Mifflin, Boston, MA

Hull, J.C. (1997) *Options, Futures, and Other Derivatives.* Prentice Hall, Upper Saddle River, NJ

Jagers, P. (1975) *Branching Processes with Biological Applications.* John Wiley and Sons, New York

Johnson, N.L. and Kotz, S. (1977) *Urn Models and Their Applications.* John Wiley and Sons, New York

Karatzas, I., and Shreve, S.E. (1991) *Brownian Motion and Stochastic Calculus.* Second Edition. Springer-Verlag, New York

Karlin, S. and Taylor, H.M. (1975) *A First Course in Stochastic Processes.* Academic Press, New York

Kelly, F. (1979) *Reversibility and Stochastic Networks.* John Wiley and Sons, New York

Kingman, J.F.C. (1993) *Poisson Processes.* Oxford U. Press

Neveu, J. (1975) *Discrete Parameter Martingales.* North Holland, Amsterdam

Resnick, S.I. (1992) *Adventures in Stochastic Processes.* Birkhauser, Boston

Revuz, D. and Yor, M. (1991) *Continuous Martingales and Brownian Motion.* Springer-Verlag, New York

Takacs, L. (1962) *Introduction to the Theory of Queues.* Oxford U Press

Answers to Selected Exercises

Answers are given to all odd numbered exercises that are not proofs.

Review of Probability

1.1. (a) 12/90, (b) 48/90. 1.3. 15/36
1.5. (a) 2/16. (b) 4/32. (c) 20/64. (d) 40/128.
1.7. (a) 90 (b) No (c) $P(C|M) = 100/170$ (d) $P(M|C) = 100/160$. 1.9. Yes.
1.11. Let $\Omega = \{1, 2, 3, 4\}$ with the four points equally likely. Let $A = \{1, 2\}$,
$B = \{2, 3\}$, and $C = \{2, 4\}$.
1.15. $P(B \cap A^c) = 0.3$, so $P(B) = 0.5$.   1.17. 7/11.
1.19. (a) 0.27 (b) 0.675   1.21. $P(E) = .17$.   1.23. 0.7
1.25. $2/n$ 1.27. (a) $p = 4/36 + (26/36)p$ so $p = 4/10$.

2.1. (a) $6 : 11/36, 5 : 9/36, 4 : 7/36, 3 : 5/36, 2 : 3/36, 1 : 1/36$.
(b) $0 : 6/36, 1 : 10/36, 2 : 8/36, 3 : 6/36, 4 : 4/36, 5 : 2/36$
2.3. $P(X = m) = \binom{m-1}{2}/\binom{15}{3}$ for $3 \leq m \leq 15$.
2.5. $P(X = i) \geq P(X = i - 1)$ if and only if $p(n - i + 1) \geq i(1 - p)$.
2.7. Poisson approximation is $e^{-1} = 0.3678$ in each case.
Exact answers (a) $(.9)^{10} = .3486$, (b) $(.98)^{50} = 0.3641$.
2.9. $(1.25)e^{-0.25} = 0.9735$ 2.11. Yes. $x^{-2}e^{-1/x}$.
2.13. (a) $x^2/4$ for $0 \leq x \leq 2$, (b) 1/4, (c) 7/16, (d) $\sqrt{2}/2$
2.15. (a) e^{-2} (b) $(\ln 2)/\lambda$ 2.17. $P(X + Y = n) = (n-1)(1-p)^n p^2$
2.19. $f_{X+Y}(z) = z/2$ for $0 < z \leq 1$, $1/2$ for $1 \leq z \leq 2$, $(3-z)/2$ for $2 \leq z < 3$
2.21. $x^2/2$ if $0 \leq x \leq 1$, $x(3-x) + 3/2$ if $1 \leq x \leq 2$, $(3-x)^2/2$ if $2 \leq x \leq 3$.
2.23. (a) $P(X + Y = z) = F(z) - F(z-1)$

3.1. $5 plus their dollar back
3.3. The probability of at least one success is 7/15, so mean is 15/7.
3.5. The mean $= 3/4$, $EX^2 = 3/5$, $\text{var}(X) = 3/80$, $\sigma(X) = \sqrt{3/80}$
3.7. (a) $c = 3/4$. (b) $EX = 1$. $EX^2 = 1.2$, $\text{var}(X) = 0.2$, $\sigma(X) = .447$
3.11. 34.   3.13. 1.   3.15. $EN_k = k/p$.   3.17. 1.05
3.19. $(50 \cdot 80 \cdot 79)/(100 \cdot 99)$ 3.21. (a) $E(X+Y) = 7$. (b) $E \max\{X, Y\} = 161/36$. (c) $E \min\{X, Y\} = 91/36$. (d) $E|X - Y| = 70/36$.
3.23. $EU^2 = 1/3$. $E(U-V)^2 = 1/6$.
3.27. (a) $EY = \mu$, $\text{var}(Y) = \sigma^2$. (c) $E e^{tY} = \exp(\mu t + \sigma^2 t^2/2)$.
3.29. (a) $\gamma_X(z) = p/(1 - (1-p)z)$.

1. Markov Chains

5.3. The stationary distribution is uniform.   9.1. No.

9.3.

(a)
	RR	RS	SR	SS
RR	.6	.4	0	0
RS	0	0	.6	.4
SR	.6	.4	0	0
SS	0	0	.3	.7

(b)
	RR	RS	SR	SS
RR	.36	.24	.24	.16
RS	.36	.24	.12	.28
SR	.36	.24	.24	.16
SS	.18	.12	.21	.49

(c) $p^2(SS, RR) + p^2(SS, SR) = .18 + .21 = .39$.

9.5.

(a)
	A	B	C
A	0	1/2	1/2
B	3/4	0	1/4
C	3/4	1/4	0

(b) At time 2, A has probability 3/4, while B and C have probability 1/8 each. The probability of B at time 3 is 13/32.

9.7. (a) 1,3,5 transient; 2,4 recurrent. (b) 3,2 transient; 1,4,5,6 recurrent.

9.9.(a)
	M	H	D	S	J	T
M	1	0	0	0	0	0
H	0	0	1/4	1/4	1/4	1/4
D	1/4	1/4	0	1/4	0	1/4
S	1/4	1/4	1/4	0	0	1/4
J	0	0	0	0	0	1
T	0	0	0	0	1	0

D, H, S are all transient; J, T, H are recurrent. (b) 2/5

9.11. (a) $\pi(1) = 1/8$, $\pi(2) = 3/8$, $\pi(3) = 4/8$.
(b) $\pi(1) = 1/5$, $\pi(2) = 1/5$, $\pi(3) = 2/5$, $\pi(4) = 1/5$.

9.13. $\pi(T) = 4/19$, $\pi(C) = 15/19$.   9.15. $\pi(P) = 4/5$, $\pi(S) = 1/5$.

9.19. 20% sunny, 40% cloudy, 40% rainy.

9.21. (i) 20/71. (ii) $E(T_2|X_0 = 2) = 71$

9.23.(a)
	1	2	3	4
1	1/3	2/3	0	0
2	1/3	0	2/3	0
3	0	1/3	0	2/3
4	0	0	1/3	2/3

(b) $\pi(1) = 1/15, \pi(2) = 2/15, \pi(3) = 4/15, \pi(4) = 8/15$

9.25.(a)
	0	1	2	3
0	0	0	0	1
1	0	0	0.8	0.2
2	0	0.8	0.2	0
3	0.8	0.2	0	0

(b) $.16/3.8 = 0.0421$.

9.27. $\pi(0) = 100/122$, $\pi(1) = 10/122$, $\pi(2) = 2/122$, $\pi(3) = 10/122$.
9.29. $1/5$.   9.31 (b) $\lim_{n\to\infty} E_x X_n = Nv/(u+v)$.

9.33.(a)

	0	1	2	3	4	5	6
0	0	1	0	0	0	0	0
1	0	1/6	5/6	0	0	0	0
2	0	0	2/6	4/6	0	0	0
3	0	0	0	3/6	3/6	0	0
4	0	0	0	0	4/6	2/6	0
5	0	0	0	0	0	5/6	1/6
6	0	0	0	0	0	0	1

(b) 14.7

9.39. (a) $0.36/0.52 = 0.6923$. (b) 0.8769. (c) 0.4154.
9.41. 8/15.   9.43. (a) 12. (b) 1/11.
9.45. (a) Squares at a distance $k = 0, 1, 2, 3$ from the boundary have probabilities $(21+k)/1456$. (b) $1456/21 = 69.333$.   9.53. $p/(1-p)$.

2. Martingales

5.1. $P_{ij}(V_{22} < V_{00}) = (i+j)/4$.   5.3. $\sigma^2/2 - \mu = 0$.

5.7.(a) $\dfrac{1}{2} \cdot \dfrac{2}{3} \cdots \dfrac{k}{k+1} \cdot \dfrac{1}{k+2} \cdot \dfrac{2}{k+3} \cdots \dfrac{n-k}{n+1} = \dfrac{k!(n-k)!}{(n+1)!}$

5.15. (c) The strong Markov property implies that $V_{x-1}, V_{x-2} - V_{x-1}, \ldots V_0 - V_1$ are independent and have the same distribution.

3. Poisson Processes

2.3. (a) 1/8. (b) $1 - e^{-1/8} = 0.1175$.   7.1. (a) e^{-1}, (b) e^{-1}
7.3. (a) 3/5. (b) 38   7.7. (a) 6 (b) 9 (c) 4/9
7.9. (a) $\exp(-\lambda_1 s - \lambda_2 t - \lambda_3(s \vee t))$. (b) U is exponential rate $\lambda_1 + \lambda_3$. V is exponential with rate $\lambda_2 + \lambda_3$. (c) No
7.11. 1.35 years.   7.13. 16 minutes
7.15. (a) $ET = 150$. (b) $4 : 1/2$, $3 : 1/4$, $2, 1 : 1/8$. (c) $ET = 50(n-1)$, $P(N = j) = 1/2^{n+1-j}$ for $2 \le j \le n$ while $P(N = 1) = P(N = 2) = 1/2^{n-1}$.
7.17. (a) $e^{-4} 4^3/3!$.

(b) $$P(N(1) = 1 | N(3) = 4) = \binom{4}{1} \left(\dfrac{1}{3}\right)^1 \left(\dfrac{2}{3}\right)^3$$

7.19. (a) $(2/5)^2$, (b) $1 - (3/5)^2$

7.21. (a) $1 - e^{-4}$. (b) 46. (c) $\binom{50}{5}(.1)^5(.9)^{45}$.
7.23. (a) $(e^{-3}3^2/2!) \cdot (e^{-1}1^3/3!)$

(b) $\binom{5}{3}\left(\frac{3}{4}\right)^3\left(\frac{1}{4}\right)^2 + 5\left(\frac{3}{4}\right)^4\left(\frac{1}{4}\right)^1 + \left(\frac{3}{4}\right)^5$

7.25. $\binom{i+j}{i}\left(\frac{\lambda_1}{\lambda_1+\lambda_2}\right)^i \left(\frac{\lambda_2}{\lambda_1+\lambda_2}\right)^j$

7.27 $(e^{-2}2^1/1!) \cdot (e^{-3}3^2/2!)$ 7.29. Mean $160K$, standard deviation $24K$.
7.31. (a) $\exp(-\lambda t[1 - g(z)])$. (b) $ES = \lambda EY_i$. (c) $ES(S-1) = \lambda EY_i(Y_i - 1) + \lambda^2(EY_i)^2$ (d) $\text{var}(S) = \lambda EY_i^2$.
7.33. (a) $N_1(t)$ and $N_2(t)$ are independent Poissons with means λpt and $\lambda(1-p)t$
(b) $P(L = k) = (1 - p)^k p$ for $k = 0, 1, 2, \ldots$.
7.35. (a) Independent Poisson with means $200(1 - p_1)(1 - p_2)$, $200p_1(1 - p_2)$, $200(1 - p_1)p_2$, and $200p_1p_2$. (b) $\hat{p}_1 = 2/3$, $\hat{p}_2 = 0.8$, 10 typos.
7.37. $P(N = n) = (1-p)^{n-1}p$ for $n = 1, 2, \ldots$, where $p = e^{-1} - e^{-2}$. (b) $1/4.3$
7.39. (a) $E(t - L) = (1 - e^{-\lambda t})/\lambda$. (b) $\to 1/\lambda$.
7.41. (a) Poisson with mean $\lambda \int_0^t (1 - F(t-s))\, ds$.

4. Continuous-time Markov Chains

8.1. (a) The state space is $\{0, 1, 2, 3, 4\}$. The transition rate matrix is

	0	1	2	3	4
0	0	0	0	0	0
1	μ	$-(\mu + 3\lambda)$	3λ	0	0
2	0	2μ	$-(2\mu + 4\lambda)$	4λ	0
3	0	0	3μ	$-(3\mu + 3\lambda)$	3λ
4	0	0	0	4μ	-4μ

(b) $11/12\lambda$.
8.3. If $0 < j \le i$ then $p_t(i,j) = e^{-\mu t}(\mu t)^{i-j}/(i-j)!$; $p_t(i,0) = 1 - \sum_{j=1}^{i} p_{ij}(t)$.
8.5. The Q matrix is:

	0	1	2	3
0	-1	0	1	0
1	2	-3	0	1
2	0	2	-2	0
3	0	0	2	-2

$\pi(0) = 2/5$, $\pi(1) = 1/5$, $\pi(2) = 3/10$, $\pi(3) = 1/10$. (b) $6/5$ per week.

8.7. (a) The transition rate matrix is given by

$$\begin{array}{c|cccc} & 0 & 1 & 2 & 12 \\ \hline 0 & -(\lambda_1+\lambda_2) & \lambda_1 & \lambda_2 & 0 \\ 1 & \mu_1 & -(\mu_1+\lambda_2) & 0 & \lambda_2 \\ 2 & \mu_2 & 0 & -(\mu_2+\lambda_1) & \lambda_1 \\ 12 & 0 & 0 & \mu_1 & -\mu_1 \end{array}$$

(b) $\pi_0 = 20/57$, $\pi_1 = 4/57$, $\pi_2 = 18/57$, $\pi_3 = 15/57$.
8.9. (a) $\pi_{12} = 1/9$, $\pi_1 = \pi_2 = 2/9$ and $\pi_0 = 4/9$. (b) The transition matrix is

$$\begin{array}{c|cccc} & 0 & 1 & 2 & 12 \\ \hline 0 & -2 & 1 & 1 & 0 \\ 1 & 3 & -5 & 0 & 2 \\ 2 & 3 & 0 & -5 & 2 \\ 12 & 0 & 3 & 3 & -6 \end{array}$$

$\pi_0 = 9/17$, $\pi_1 = \pi_2 = 3/17$, $\pi_{12} = 2/17$.   8.11. (a) 6/9. (b) 4/9.
8.13. $\pi_i = c\lambda_i/\mu_i$ for $i = 1, 2, 3$ where c makes the π_i sum to 1.
8.15. (a) $\pi_0 = 1/7$, $\pi_1 = 2/7$, $\pi_2 = 4/7$. (b) $60/7 = 8.57$ per hour.
8.17. $\pi(3) = (2/3)^3$, $\pi(2) = 3(2/3)^2(1/3)$, $\pi(1) = 3(1/3)^2(2/3)$, $\pi(0) = (1/3)^3$.
8.19. (a) X_t is a birth and death chain with birth rates $\lambda_2 = 1/2$, $\lambda_1 = \lambda_0 = 1$, death rates $\mu_1 = 1/20$, $\mu_2 = 2/10$ and $\mu_3 = 3/20$.

$$\pi(3) = \frac{2000}{2663}, \quad \pi(2) = \frac{600}{2663}, \quad \pi(1) = \frac{60}{2663}, \quad \pi(0) = \frac{3}{2663}$$

(b) $63/2663$ of the time. (c) $7260/2663 = 2.726$.
8.21. (b) About .0386 customers per hour.   8.23. (a) 4/157. (b) 306/157
8.25. (b) Poisson with mean λ/μ.   8.27. (i) $q(n, n+1) = \lambda$ for $n \geq 0$; $q(n, n-1) = \mu$ for $n \geq 1$; $q(n, -1) = \alpha$ for $n \geq 1$; $q(-1, 0) = \beta$.
8.29. $\pi_n = \pi_2(\lambda/\mu)^{n-2}$ for $n \geq 2$, $\pi_a = c\mu_b/\lambda$, $\pi_b = c\mu_a/\lambda$. $\pi_0 = 2c\mu_a\mu_b/\lambda^2$.
8.33. $c_N(\lambda_1/\mu_1)^{n_1}(\lambda_2/\mu_2)^{n_2}$.   8.35. $\pi(n) = c_s e^{-\lambda/\mu}(\lambda/\mu)^n/n!$
8.37. The stationary distribution is uniform over the set of possibilities.
8.39. Independent shifted geometrics with failure probabilities 3/4, 3/5, 4/6.
8.41. $r_i = \lambda_i + \lambda_{i-1}p(i-1, i) + \cdots + \lambda_1 p(1, 2) \cdots p(i-1, i)$. For stability we must have $r_i < \mu_i$ for all i.   8.43 $.5r_1 = 7.921$, $.4r_2 = 7.7892$, $.7r_3 = 8.2894$.
8.45. 252/277   8.47. All the states are equally likely.

5. Renewal Theory

5.1. (a) T_n is Poisson with mean $n\mu$. (b) $P(N(t) \geq n) = \sum_{k \leq t} e^{-n\mu} \frac{(n\mu)^k}{k!}$.

5.3. $$P(N(t) = n) = e^{-\lambda t} \frac{(\lambda t)^{2n}}{(2n)!} \left(1 + \sum_{k=1}^{m-1} \frac{(\lambda t)^k}{(2n+1)\cdots(2n+k)}\right)$$

5.7. (a) The Laplace transform

$$\varphi(\theta) = \frac{3+2\theta}{(1+\theta)(3+\theta)}$$

5.9. 2/9.   5.11. 2/3.   5.13. $102/7.5 = 13.6$ dollars per minute.
5.15. (a) $\tau/(\tau + 1/\lambda)$. (b) $1/(\lambda\tau + 1)$.   5.17. $\mu/(\mu + \lambda)$
5.19. The limiting fractions are 3/12, 4/12, and 5/12.
5.21. (a) $e^{-0.3}$. (b) $Eu_i = 2(e^{0.3} - 1) = .70$.

5.23.(a) $\quad \dfrac{4 - 2e^{-\lambda T}}{\frac{1}{\lambda}(1 - e^{-\lambda T})} = 2\lambda\left(1 + \dfrac{1}{1 - e^{-\lambda T}}\right)$

5.25. (a) $\pi_0 = 1/7$, $\pi_1 = 2/7$, $\pi_2 = 4/7$. (b) 10 minutes. (c) 4.
5.27. (a) $\pi(4) = 81/203$, $\pi(3) = 54/203$, $\pi(2) = 36/203$, $\pi(1) = 24/203$, $\pi(0) = 8/203$ (b) .59.   5.29. 1/4.

5.31 $\qquad \lambda e^{-\lambda t} \cdot \dfrac{1}{5} \sum_{m=1}^{5} \dfrac{(\lambda t)^{m-1}}{(m-1)!}$

5.35. $(1 - F(x + s))/(1 - F(s))$.   5.37. (a) 2/7. (b) $\pi_0 = 6/21$, $\pi_1 = 5/21$, $\pi_2 = 4/21$, $\pi_3 = 3/21$, $\pi_4 = 2/21$, $\pi_5 = 1/21$

6. Brownian Motion

6.5. The joint distribution of (Y_t, B_t) is bivariate normal, so the conditional distribution is normal. The mean $E(Y_t|B_t = z) = t \cdot \frac{z}{2}$. $E(Y_t^2|B_t = z) = t^3/6$.
6.17. The density $p_t(x, y) - p_t(x, -y)$.   6.21. $P(R_t = u) = \dfrac{1}{\pi\mu} \cdot \sqrt{\dfrac{t}{u-t}}$.
6.23. $b = (a^2 + 3a)/(a - 1)$. When $a = 2$, $b = 10$.   6.31. (i) The equation is $\alpha\theta^2 + \beta\theta + \gamma = 0$ where $\alpha = 1/2$, $\beta = \mu/\sigma$, and $\gamma = -\lambda$ so the roots are $-\beta \pm (\beta^2 - 4\alpha\gamma)/2\alpha$.   6.35. 61/15.
6.39. Only one probability distribution $p_1 = .3$, $p_2 = .2$, $p_3 = .5$ makes the two stocks martingales, so the option is worth 6.

Index

absorbing state 33, 41
alternating renewal process 219
American call option 263
aperiodic 50
arbitrage opportunity 258

Backgammon 252
backwards random walk 109
ballot theorem 120
barbershop 173, 203, 224
Bayes formula 4
Bernouli-Laplace model of diffusion 95
Binomial distribution
    mean and variance 23
    sums of 15
Binomial theorem 15
birth and death chains 61, 63, 97, 123, 164, 173
Black-Scholes model 103, 261
branching processes 32, 77, 103, 124
brother-sister mating 90, 121
Brownian bridge 245
Brownian motion
    covariance 244
    definition 242
    exit distribution 251
    exit time 253
    law of iterated logarithm 269
    Markov property 246
    nondifferentiability 266
    quadratic variation 266
    scaling relation 244
    strong law 265

Chapman-Kolmogorov equation 36, 159
closed set for Markov chain 45
communicates with 42
complement of a set 2
compound Poisson process 137

conditional expectation 100
conditional probability 3
continuous distribution 9
convergence theorem for Markov chains 54, 76, 82
coupon collector's problem 95
counter processes 210
Cramer's estimate of ruin 119

decomposition theorem 45
delayed renewal processes 234
density function 9
detailed balance condition 61, 172, 205
discrete renewal theory 240
disjoint sets 2
distribution function 10
doubly stochastic 59
doubling strategy 110
duration of fair games 69, 117, 122

Ehrenfest chain 29, 49, 62, 94
European call option 257
event 1
expected value 18
experiment 1
exponential distribution 10, 11, 126
    mean and variance 21, 22
    sums of 16, 129
exploding Markov chain 162
exponential martingale 107, 254
exponential races 127

feed-forward queues 206

gamma distribution 16, 22
gambler's ruin 28, 38, 41, 66, 107, 115, 116, 122, 123
Gaussian process 244
generating function 24

geometric distribution 8, 19
GI/G/1 queue 123, 221
hitting probabilities 124
hitting times 124

independent events 4, 5
independent increments 132, 246
independent random variables 13
indicator function 100
inspection paradox 230
intersection of sets 2
inventory chain 32, 49
irreducible Markov chain 45, 170

joint density function 12

Kolmogorov's
    backward equation 165
    cycle condition 205
    forward equation 166
kth moment 18

lack of memory property 12, 127
landscape dynamics 94
left-continuous random walk 117
Lévy's interpolation construction 265
library chain 96
likelihood ratios 108, 121
Little's formula 224
logistic model 204
Lyapunov function 123

machine repair 31, 49, 56, 63, 174, 195, 199, 209
manufacturing system 195, 198
marginal density 13
marginal distribution 12
Markov chain 28, 158
martingale defined 102, 106
matching pennies 67
mean 18
median 17
M/G/1 queue 80, 226, 227
M/G/∞ queue 151
migration process 191, 193

M/M/1 queue 176, 190, 210, 224, 227
M/M/s queue 160, 161, 175, 179, 181
M/M/∞ queue 180, 192
moment generating function 23
multiplication rule 3

nonhomogeneous Poisson process 136, 149, 151
normal distribution 10
null recurrent 75

option pricing 108, 257
Ornstein-Uhlenbeck process 245

pairwise independent 5
PASTA 226
pedestrian lemma 44
period of a state 50
Poisson approximation to Binomial 9, 135
Poisson distribution 9, 131
    mean and variance 19, 131
    m.g.f. 24, 131
    sums of 14, 131
Poisson process 130, 145, 160, 167
Pollaczek-Khintchine 228
Polya's urn scheme 109
positive part 32, 109
positive recurrent 75
probability 2
probability function 8
probability of ruin in a favorable game 118, 254

queues in series 206

random variable 8
random walks 73, 92, 125
random walks on graphs 64, 65, 97
recurrent state 41
reflection principle 247
renewal process 209
    limiting behavior 214, 233
residual life 231
roulette 68

sample space 1
simple exclusion process 206
simple random walk 114
spatial Poisson processes 145
stationary distribution 53, 55, 170
stock prices 103, 121
stopping theorem 114, 124, 125, 250
stopping times 40, 112, 212
strong law of large numbers 214
strong law for
    Brownian motion 265
    Markov chains 56, 87
    Poisson processes 214
    renewal processes 214
    renewal reward processes 216
strong Markov property 40, 150
submartingale 102
supermartingale 102
superposition of Poisson processes 141

thinning of Poisson processes 141
traffic flow 148, 210
transient state 41
transition probability 28, 159
two station queues 185, 187, 202

unfair fair game 121
uniform distribution 10, 11
    mean and variance 20, 22
    sums of 16
union 2

Wald's equation 116, 212
waiting times for coin patterns 70
waiting times in queues 225
weather chain 30, 37, 39, 43, 54, 61, 171
Wright-Fisher model 33, 68, 94, 104, 121

Yule process 163, 167

Springer Texts in Statistics *(continued from page ii)*

Madansky: Prescriptions for Working Statisticians
McPherson: Applying and Interpreting Statistics: A Comprehensive Guide, Second Edition
Mueller: Basic Principles of Structural Equation Modeling: An Introduction to LISREL and EQS
Nguyen and Rogers: Fundamentals of Mathematical Statistics: Volume I: Probability for Statistics
Nguyen and Rogers: Fundamentals of Mathematical Statistics: Volume II: Statistical Inference
Noether: Introduction to Statistics: The Nonparametric Way
Nolan and Speed: Stat Labs: Mathematical Statistics Through Applications
Peters: Counting for Something: Statistical Principles and Personalities
Pfeiffer: Probability for Applications
Pitman: Probability
Rawlings, Pantula and Dickey: Applied Regression Analysis
Robert: The Bayesian Choice: From Decision-Theoretic Foundations to Computational Implementation, Second Edition
Robert and Casella: Monte Carlo Statistical Methods
Rose and Smith: Mathematical Statistics with *Mathematica*
Santner and Duffy: The Statistical Analysis of Discrete Data
Saville and Wood: Statistical Methods: The Geometric Approach
Sen and Srivastava: Regression Analysis: Theory, Methods, and Applications
Shao: Mathematical Statistics, Second Edition
Shorack: Probability for Statisticians
Shumway and Stoffer: Time Series Analysis and Its Applications
Simonoff: Analyzing Categorical Data
Terrell: Mathematical Statistics: A Unified Introduction
Timm: Applied Multivariate Analysis
Toutenburg: Statistical Analysis of Designed Experiments, Second Edition
Whittle: Probability via Expectation, Fourth Edition
Zacks: Introduction to Reliability Analysis: Probability Models and Statistical Methods